REACHING CHILDREN IN WAR
SUDAN UGANDA AND MOZAMBIQUE

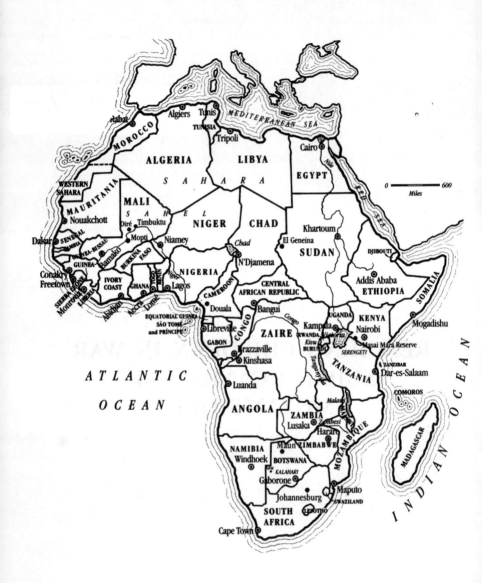

REACHING CHILDREN IN WAR SUDAN UGANDA AND MOZAMBIQUE

Cole P. Dodge
Magne Raundalen

Sigma Forlag, Bergen, Norway
Scandinavian Institute of African Studies
Uppsala, Sweden

Extracts may be reproduced by the press or by non-profit organizations with acknowledgement to the authors and publishers.

© Copyright 1991 Cole P. Dodge and Magne Raundalen

ISBN # 82-90373-61-9
ISBN # 91-7106-319-6
   Published by Sigma Forlag
   in cooperation with
   Nordiska Afrikainstitutet
   (The Scandinavian Institute of African Studies)

The views expressed by the authors do not necessarily represent the views of the organizations with which they are affiliated. The opinions expressed by Cole P. Dodge are those of the author and do not necessarily reflect the policy or views of the United Nations Children's Fund.

Cover photo credit UNICEF/Yann Gamblin
Back cover photo Shehzad Noorani

This book was typeset in Times Roman. Desktop publishing by Tauhidur Rashid.
Printed in Bangladesh by BRAC Printers, 66 Mohakhali C/A, Dhaka 1212

Contents

Glossary		vii
Acknowledgement		ix
Foreword	*James P. Grant*	xii
Introduction	Personal reflections on the new rights for children in war *Lisbet Palme*	1
Chapter 1	National and societal implications of war on children *Cole P. Dodge*	7
Chapter 2	War experiences and psychological impact on children *Magne Raundalen and Atle Dyregrov*	21
Chapter 3	Fleeing the war : street boys in Khartoum and Maputo *Cole P. Dodge and Magne Raundalen*	39
Chapter 4	Child soldiers of Uganda and Mozambique *Cole P. Dodge*	51
Chapter 5	Internally displaced: a silent majority under stress *Rune Stuvland and Cole P Dodge*	59
Chapter 6	Corridors of peace across the lines of civil war in Uganda and Sudan *Cole P. Dodge*	71

Contents

Chapter 7	Helping the child *Magne Raundalen and Atle Dyregrov*	89
Chapter 8	Research challenges in practical perspective *Magne Raundalen and Cole P. Dodge*	103
Chapter 9	Rights and hopes for children in war *Magne Raundalen and Cole P. Dodge*	113
Annex	Convention on the rights of the child	123

GLOSSARY

AMREF	African Medical Research Foundation
Bairros	Community
BBC	British Broadcasting Corporation
CNN	Cable News Network
Curandeiros	Spiritual leaders
Deslocados	Displaced
DSM III	Diagnostic Statistics Manual Third Revision
FAO	Food and Agriculture Organization
Frelimo	Front for the Liberation of Mozambique
FUNA	Former Uganda National Army
GNP	Gross National Product
ICRC	International Committee of Red Cross
IMF	International Monetary Fund
IMR	Infant Mortality Rate
Intifadah	Uprising
Merissa	Sorghum beer
MNR	Mozambique National Resistance
MSF	Médecins Sans Frontières
NGO	Non-Governmental Organization
NRA	National Resistance Army
NRM	National Resistance Movement
OLS	Operation Lifeline Sudan
ORS	Oral Rehydration Salts
PTSD	Post Traumatic Stress Disorder
Regulos	Local chiefs
Renamo	Mozambican National Resistance
SABAH	Street Boys Centre in Khartoum
SADCC	South African Development Coordination Conference
SCF	Save the Children Fund
Shamassa	Street boys
Sharia	Islamic Law
SKI	Street Kids Incorporated
SPLA	Sudan Peoples Liberation Army
SPLM	Sudan Peoples Liberation Movement
SRRA	Sudanese Relief and Rehabilitation Agency
UN	United Nations
UNDP	United Nations Development Programme
UNHCR	United Nations High Commission for Refugees
UNICEF	United Nations Children's Fund
UNLA	Uganda National Liberation Army
WHO	World Health Organization
WFP	World Food Programme

GLOSSARY

AMREF	African Medical Research Foundation
Barrio	Community
BBC	British Broadcasting Corporation
CNN	Cable News Network
Curandeiros	Spiritual leaders
Deslocados	Displaced
DSM III	Diagnostic Statistics Manual Third Revision
FAO	Food and Agriculture Organization
Frelimo	Front for the Liberation of Mozambique
FUNA	Former Uganda National Army
GNP	Gross National Product
ICRC	International Committee of Red Cross
IMF	International Monetary Fund
IMR	Infant Mortality Rate
Jandudu	Uprising
Jerusa	Sorghum beer
MNR	Mozambique National Resistance
MSF	Médecins sans Frontières
NGO	Non-governmental Organization
NRA	National Resistance Army
NRM	National Resistance Movement
OLS	Operation Lifeline Sudan
ORS	Oral Rehydration Salts
PTSD	Post Traumatic Stress Disorder
Regulos	Local chiefs
Renamo	Mozambican National Resistance
SABAH	Street Boys Centre in Khartoum
SADCC	South African Development Coordination Conference
SCF	Save the Children Fund
Shamassa	Street boys
Sharia	Islamic Law
SKI	Street Kids Incorporated
SPLA	Sudan People's Liberation Army
SPLM	Sudan People's Liberation Movement
SRRA	Sudanese Relief and Rehabilitation Agency
UN	United Nations
UNDP	United Nations Development Programme
UNHCR	United Nations High Commission for Refugees
UNICEF	United Nations Children's Fund
UNLA	Uganda National Liberation Army
WHO	World Health Organization
WFP	World Food Programme

ACKNOWLEDGEMENTS

The children of Mozambique, the Sudan (north and south, especially the south) and Uganda shared their aspirations with us and our research assistants. We hope this has been worthwhile and that their experience will give some indication to the next steps to reach children and the child in war.

We met through our work for UNICEF; Magne Raundalen, a practicing child psychologist, was Chairman of the UNICEF Committee in Norway and Cole Dodge, an anthropologist, was fund raising for children in Uganda. Over dinner, we agreed to work together on the issue of how war affected children in Africa. Lisbeth Palme received us at her cottage on Färö in the summer of 1989 and spent a week with us in Dhaka in February 1990. As we struggled with the final stages of writing, she again contributed in January 1991. Her personal commitment to reaching children of war in Africa was an inspiration. While the book does not deal with each country separately, or flow from beginning to end, we feel it is of value for the insights which the individual chapters contain. Our respective jobs demand full time attention, therefore, we have had to fit the makings of this book into vacations and weekends.

Numerous young researchers, interviewers, relief workers and insightful individuals have helped. In Khartoum: Munir el Safi, Director of Sabah, Sulieman el Amien who directed SKI, Peter Dalglish and Blanka el Khalifa who cared about street children, Abdul Mohammed and Priscilla Kuch who worked for the displaced. Terje Halvorsen in Khartoum and in Mozambique, Nils Vogt, NORAD Representative, were always supportive. Eduarda Pereira, Tora Synøve Raundalen and Rune Stuvland interviewed children. Esterela Isabel, Issufo Baira Mamudozouira, Edwards Pereira and Iris Aslaksen from Norwegian Red Cross helped us. And in Uganda, people like Senteza Kajubi who took the idea of essays and created MINDACROSS and Violet Mugisha who has kept the child-to-child programme going. The task of entering the text into various computers was effectively executed by Rune Stuvland who also contributed his understanding. Kjersti Johansen, secretary, Centre for Crisis Psychology, typed the original drafts. In Dhaka, Marilyn Dodge entered, corrected and edited the text and Rashid Tauhidur formatted and printed the final copy.

This project would not have been possible without the understanding of our families and employers, United Nations Children's Fund (UNICEF), the University of Bergen and the Centre for Crisis Psychology. The Norwegian Red Cross supported some of the research in Mozambique. Katrine Kloster, the Norwegian psychologist and philanthropist, supported Research and Action for Children at the Centre for Crisis Psychology, which in the final stages, enabled us to complete the manuscript.

FOREWORD

James P. Grant

Executive Director
United Nations Children's Fund

Perhaps the only way to fully protect children from the effects of war is to achieve and maintain peace.

While armed conflicts still rage through adult societies, it is the children who suffer most in the present, and all too many lose their futures to the ravages of war.

This book presents a valuable look at what it is that children suffer in the wake of adult violence, and at some of the measures that can be taken to help them recover and cope.

During war, children still need food. They need health care, clean water, shelter and clothing. They still need an education, and help in understanding what is going on around them.

Unfortunately, children need more and more psychological help in coping with the traumas of war, as they are increasingly exposed to the brutal front lines of confrontation. It is a sad commentary on our civilization that the percentage of civilian war casualties has risen dramatically in the 20th century. This book documents an increase from 10 percent in World War I to approximately 80 percent in conflicts since the Second World War. And when civilians are the victims, it is as usual the most vulnerable -- especially children and women -- who suffer a disproportionate share of the burden.

Some of the damage done to children during war can never be repaired. Children are killed in war. And there are damages to so many who survive which are also, tragically, beyond full repair. A maimed leg will never carry a child through the games of childhood, nor will it, when that child matures, contribute to her productive part in society. So too, many traumatized psyches will never fully erase the effects of violent horrors a child has witnessed and experienced.

Immediate actions are needed wherever children are threatened by the effects of war. It was the mounting recognition of this fact -- both legally and as a collective ethic -- that moved the United Nations to take a most reasonable but unprecedented action during the Gulf Crisis in the winter of 1991. At the height of military activities, a humanitarian mission was sent into the heart of the conflict, to Baghdad, carrying emergency medical supplies for children and women. The joint WHO/UNICEF team received safe passage from all of the warring parties.

Foreword

How was it that the massive military powers which had amassed in that conflict would hold back fire and stay their tactics in order to help safeguard children? How could children -- the least powerful and most vulnerable among us -- cause such a rational pause in the heat of military aggression?

When the Secretary-General of the United Nations announced the departure of the convoy, he emphasized that the mission was in keeping with provisions of both the Convention on the Rights of the Child and the Plan of Action of the World Summit for Children. Governments had, in September of 1990, committed themselves in these two landmark "documents" to meeting the basic needs of children in times of war and to protecting children from the effects of war.

The specifics of these agreements among governments of the world regarding adult obligations to children in times of war are discussed in greater detail in this book than in these few introductory words. But I will note here that it marks a great advance that we can now demand peaceful activities in times of war. They can be demanded in the name of children, and in accordance with agreements at the highest level of governments.

And I will note here that perhaps no other issue **but** children could have caused, at this stage in history, warring parties to agree to synchronize constructive activities. It is an innovation which as been building. It started in El Salvador, where, since 1985, civil war has been interrupted each year with three "days of tranquility" during which children are immunized against the main child killing diseases. That practice was emulated with similar cease-fires first in Lebanon, then in Uganda. In the midst of heated conflict, days were set aside when the only "shots" were those that protected young lives with vaccine.

The concept evolved quickly. In 1989, Operation Lifeline Sudan (OLS) negotiated "corridors of tranquility" for the safe passage of food supplies through civil conflict to the famine-starved south of Sudan. A repetition of the previous year's tragic loss of 250,000 lives to starvation -- the majority of them children and women -- was thus prevented.

The spectre of innocent children suffering and dying as a result of adult aggression does, indeed, warrant a radical change in the way we act. Even opposing factions in conflict, regardless of their differences on other issues, can all be moved by the plight of children. These days and corridors of tranquility bring great hope that the experience of cooperation around a peaceful, constructive activity by otherwise warring parties will help lead to a far broader peace.

The same steps that we must take to care for those children who currently suffer the tragedies and indignities of war will take us a long way, I believe, toward the ultimate goal of protecting children from the effects of war through achieving and sustaining peace. We must demand that resources and actions of our societies are focussed where they are needed, constructively, and not on the purposes of aggressive solutions to conflict.

As we approach the end of this 20th century, there is a great polarization in how we, as societies, react to children in war. On one side, armed conflict is turning more and more on civilians, and it is children who suffer, in great disproportion, the brunt of the burden. On the other side, society is awakening to its obligations to children as evidenced in the provisions of the Convention, the promise of the Summit, and the landmark corridors of peace through the Gulf Conflict.

In which direction will we have swayed by the turn of the century? Which force will define our era? Can we learn what children need as a result of the aggression they have suffered, and ensure that those needs are met? Can we take all of the necessary steps to focus a first call on the resources of our societies for the survival, protection and development of children, even in the face of grave conflicts? For the children --and the future -- of the world, **working together**, I think we can. In this way too, children are our greatest hope.

New York, September 1991

INTRODUCTION

PERSONAL REFLECTIONS ON THE NEW RIGHTS FOR CHILDREN IN WAR

Lisbeth Palme

Chairperson National Committee for UNICEF, Sweden
Chairperson, UNICEF Executive Board 1990-91

When I face the suffering child tormented by the madness of war, I feel helpless. Helplessness creates anxiety. Overwhelmed by inner fear, we have different ways of thinking and acting which can provide relief: we may close our eyes, turn our backs and deny that these war-traumatized children exist, or we can determine to face reality and open our minds and hearts and feel the pain.

The reason why it is so painful is that we are at the edge of a precipice, we realize that hell exists, out there we see the atrocities on the battlefield and inside ourselves we are confronted with the human capacity for violence and war. But I feel I have no choice. I cannot deny or turn my back. I cannot say that I do not know.

I know the suffering child: the face, the scars, the wounds, the empty eyes, their swollen stomachs, their infected and blinded eyes: all these terrible pictures release an immense reaction within me.

It has been my privilege during the years to see, meet and collaborate with dedicated people from many different organizations and nations who set up to help the suffering children of the world. I have seen and felt the enormous capacity for empathy and the desire to help. I know that individuals and their organizations have contributed to a growing awareness of the sufferings of children and out of our joint commitment grows the safety net of the United Nations. When the United Nations drafted and approved the Convention on the Rights of the Child, this gave us all a real opportunity and mandate to help children in war. For many of the children I have met and talked with, the Convention takes on a very meaningful reality. For the children I met in Mozambique, article 39 in the Convention is extremely important.

Article 39 states:

"State Parties shall take all appropriate measures to promote physical and psychological recovery and social re-integration of a child victim of: any form of neglect, exploitation, or abuse; torture or any form of cruel, inhuman or degrading

treatment or punishment; or armed conflict. Such recovery and reintegration shall take place in an environment which fosters the health, self-respect and dignity of the child."

These new rights to follow-up care after traumatic experiences are not only limited to physical and psychological recovery, but social integration is also seen as essential. The phrase "*all appropriate measures*" is not very specific, but we must remind ourselves that today's children are our future.

I had the honour to attend The World Summit for Children, the largest ever gathering of heads of governments, held on September 30th 1990 at the United Nations in New York. In spite of the shadow cast by the Gulf Crisis, everyone who attended the meeting was struck by the importance of focusing on childrens' needs. The changes and improvements for children will not come overnight. But our starting point is better than ever because we now have a common frame of reference, stated in the articles of the Convention and in the Declaration and Plan of Action resulting from the Summit for Children. This frame of reference is not only shared by 71 heads of governments but also by the entire United Nations system and by many non-governmental organizations. Personally I regard these commitments as considerable steps in the right direction. They are very special and positive signs revealing that in this age human beings cannot and will not tolerate complacency in the face of children being killed, traumatized or wounded in the wars that rage around the world today.

My childhood

I was a child during the Second World War. While the war raged all around us, we children prayed that Hitler would not come to Sweden during the night. We knitted blankets, stockings and gloves for the children in the war area. From the other Nordic countries and from the Baltic states, families and children came to us as refugees. From Finland, many children without parents arrived. Now, fifty years later, we still hear from them. They have carried grief and pain within themselves all these years as a result of that long-ago separation from their parents. Yet their parents sent them away because they saw this as the only possibility to save them from the war's deadly reality.

The children from Finland, not understanding the language in their new country, could not express themselves, which greatly complicated their initial adjustment. They had no idea where they would be sent and who their caretakers would be. They had no idea when they would return home and they did not know if their parents would be alive when they finally returned.

I will never forget the small Jewish children who came to our country, rescued from the Holocaust of the German gas chambers, or the tense and jittery little girl who came from bomb-blitzed London. I continued to say my nightly prayer: "*God be Good, let there be peace. God please, do not let them come tonight.*" I can still envisage the gaunt bodies piled in heaps at the Nazi concentration camps.

When the war ended, I was 14 years old and, along with my friends, felt enormous relief and confidence that there would be no more war. In my future expectations, the negotiation table would be the only solution! During the decades that followed, I have often felt both deeply disappointed and frustrated because of all the wars that still occur all over the world between and within countries.

Weapons are very destructive to children

We can all develop in our minds the picture of the little napalm-maimed Vietnamese girl running naked through the street. On TV, we saw the inhabitants of the villages of Kurdistan who died with their children in their laps under the yellow sky of gas from the Iraqi forces. We can also recall the African children, armless or legless victims of warfare, having walked among mines or, with a child's innocent curiosity, having picked up a mine which they thought was a plaything. Toys constructed by cynical brains and distributed by cynical military strategists.

Nuclear weapons, the most terrible threat in our time, thrived for decades in a climate of cold war, becoming an obstacle to peaceful relations between states. The arms race, the tension between the superpowers, the escalating resources required to develop these weapons and counter-defensive systems from year to year, left development needs unmet year after year in Asia, Africa and Latin America. Espionage, terrorism and disinformation were assigned the highest priority in the insane years of the arms race, while the environment was steadily degraded.

From 1988 to the autumn of 1990, the years when *"peace broke out,"* we all felt optimism and hope on behalf of the world's children. Since then, shadows have appeared as events reached crisis proportions in the Gulf Crisis and developments unfolded in Eastern Europe.

How can we get started?

The suffering of children in war is a major concern for health workers all over the world. War-induced childhood trauma or other extreme experiences can affect the individual and society for decades. This was first substantiated by research focusing on the long-term after-effects on adult Holocaust victims from the German concentration camps. Furthermore, recent research from Finland has revealed that a large number of adults have lived their whole lives with tremendous psychological problems because of the trauma they experienced during the Winter War in 1939. Children evacuated to Sweden and thus separated from their families, continued to suffer as adults.

How massive is the denial of children's suffering? From time to time you may hear even educated and well informed people questioning whether war-affected children are traumatized or otherwise psychologically disturbed. Since several of these statements come from field workers in war-tormented countries like the Sudan and Mozambique, they cannot simply be pushed aside. If we examine these statements, their content is characterized by the following features:

- children cope under all conditions and they can withstand unbelievable amounts of stress,
- children do not have the capacity to fully understand war situations and are thus protected, and,
- as long as they are not separated from close adults they are not significantly affected by war experiences.

Adults, parents included, systematically underestimate children's experiences and sufferings in times of war and in other extreme situations. There is no reason to doubt that children cope and that many of them cope very well even when facing the reality of war. The relevant question, however, is how children cope and at what psychological cost? The relationship between short- and long-term after-effects of dealing with the threat of war must not be underestimated. Similarly, the relationship between immediate coping mechanisms and long-term after-effects is pertinent. What, for example, is the "price" exacted for staying seemingly unaffected without talking to anybody and without showing any emotional signs of stress?

I believe that children suffer very much in war. Many children experience direct threat. Many are physically wounded. In a survey from Chimoio in Mozambique, published for the first time in this book, it was found that 75 percent of the children surveyed had experienced a war situation where they thought they would be killed. Many children see and experience homes turned to ruins and a mother and/or father lost or missing because of war. A survey smuggled out of southern Sudan in March 1989 showed that 20 percent of the school children surveyed lost their mother because of the war and 22 percent, their father. The overlapping was very small despite some 40 percent of the sample having lost one parent.

In times of war, children may also lose siblings or others close to them. Thus many of the children build in their minds the fear that their turn might come.

- To reach children in war we need organizations active around the world like UNICEF, the Red Cross, Save the Children and the many other professional and voluntary agencies who help.
- We need new ethics allowing these organizations to reach out to children across the front line.
- We need resources to reunite children and their families when separated.
- We need knowledgeable and skilled persons to help children overcome and work through their war-related traumas, to help them avoid a life-time's grief and suffering.

The possibilities are there -- that is what *Reaching Children in War* is about. We cannot indulge in wishful thinking. We are determined to make a reality of the Convention, the Declaration and the Plan of Action from the Summit. The children cannot wait. They need support now.

CHAPTER 1

NATIONAL AND SOCIETAL IMPLICATIONS OF WAR ON CHILDREN

Cole P. Dodge

Budget allocations

War is expensive. Not only to human life but also to governmental budgets, particularly the soft social sectors which are so important to children. Many countries allocate and invest much greater amounts of money in their war efforts than in health. Figure 1 provides a quick reference for comparison.

It is interesting to note that while military expenditure has increased two and a half times since 1960 worldwide, the increase in the developing countries has been seven-fold.[1] Military spending in Africa went from $8.5 billion in 1970 to over $15 billion in 1987 [2] while the allocation for health stood at $3.8 billion.[1]

During former President Obote's second reign in Uganda, actual expenditure for the army, police and internal security was 44 percent of the budget in 1981. In subsequent years, this was reduced to around one quarter on published budgets, but many observers felt it was higher.[3] Similarly the Sudan's civil war in the south cost the government an additional $1 million per day, according to unofficial sources, while allocations for health sank from $1 to 50 cents per capita after devaluation of the currency in 1987. This increased the percentage of military expenditure to around 7.5 percent of the Gross National Product (GNP) by the late 1980s.

Reductions in health budgets in the Sudan s southern regions disrupted medical services even more than the war in the government-controlled garrison towns. Consequently, doctors, nurses and administrators who struggled to keep medical facilities open in the government-held towns of southern Sudan had their salaries and supplies reduced. Their only recourse was to descend on Khartoum where they approached non-governmental organizations (NGOs), bilaterals and United Nations (UN) agencies appealing for emergency medicines to run their hospitals and clinics.[4] With the exception of Juba, Yei and Yambio in Equatoria and Raja in Bahr el Ghazal, all other hospitals and health services were severely impaired or closed completely.

Figure 1   Comparative resources allocations
            (US $ millions)

	GNP	Military	GNP%	Health	GNP%
Ethiopia	4742	442	9.3%	66	1.4%
Sudan	8680	283	3.3%	20	.2%
Uganda	6402	68	1.1%	10	.2%
Mozambique	5667	272	4.8%	45	.8%
Angola	6970	988	14.2%	84	1.2%
Somalia	1759	168	10%	11	.6%
Chad	604	63	10.4%	4	.6%

Source : Compiled from *World Military and Social Expenditures 1987-88*, World Priorities, Washington D.C. 1987

Impact on people

There is no doubt that war has been devastating for the people of countries fighting civil wars. The death toll has been high, as evidenced by mass graves in southern Sudan or the human remains found in the villages of the Luwero Triangle of Uganda. Was it 100,000 or more who lost their lives in Luwero between 1981 and 1985? Was the toll 250,000 in southern Sudan in 1987-88? We shall never know the exact number but, judging by casualty counts, we do know that civilians have increasingly become the target of war: the First World War recorded 10 percent; the Second World War 50 percent; and, for all subsequent wars, around 80 percent of casualties were civilians.[1] See Figure 2 for a comparison of civilian and military deaths in five civil war-torn countries of east and southern Africa which averaged an incredible 92 percent civilian casualty rate. The casualty count approach to measuring the impact of war on health has to be seen in a wider context since it is not direct fighting which claims so many lives. The negative public health impact induced by war has far wider ramifications than just the conflict. Countries torn by civil war rank among the highest for infant mortality rate (IMR): Afghanistan is number one: Mozambique, number two; Angola, number four: Ethiopia, seven, and Somalia, thirteen, to give a few examples.[5] Sudan ranks twenty-fifth and Uganda thirty-first. However, as we will see later, much higher IMR occurs in times or areas of conflict and published IMR for many countries engaged in war may greatly underestimate mortality or be based on pre-war estimates.

War impacts on health and medical service delivery in a variety of ways ranging from the outright destruction of physical facilities to the flight of doctors, displacement of people, decline in investment for health services and shortages of supplies and equipment. Decline in agriculture, transport services, commerce

Panel 1

"The memories of that period will never leave me. I arrived in Abeyei in 1988 on a short shuttle hop from our base in Kadugli. The agencies were airlifting food to this isolated village on the northern side of the river which divides the country. Even though a feeding station had been established, I encountered a devastating scene of a family with four dying daughters.

"They were all emaciated to the point where they could not walk. Almost naked, the four sisters were carried, one by one, from a dark hut into the glaring sunlight. Their pictures were broadcast to millions around the world on the Cable News Network. Their story was typical of those southerners forced to flee the war but trapped by the seasonal rains which isolated this little village. Without food, Abeyei was off limits to relief agencies because of fighting and government restrictions."

As the video cameras recorded their suffering, one government official told the interviewer: *"See what the rebels are doing to their own people?"* A few weeks later, I was in the heart of southern Sudan where a rebel spokesman pointed to the hungry and said, *"See what the government is doing to its citizens?"*

Relief Worker, Sudan, 1988.

and shortage of money and the consequent impact on nutritional status and disruption of public health programmes result in susceptibility to disease and epidemics.

Figure 2   Civilian vs military deaths in selected African countries, 1970-87

	Years	No. Civ.	%	No. Mil	%	Total
Angola Civil	1975-87	200,000	94%	13,000	6%	213,000
Ethiopia Eritrea	1974-87	500,000	92%	46,000	8%	546,000
Ogaden	1972-80	15,000	42%	21,000	58%	36,000
	Sub-Total	515,000	88%	67,000	12%	582,000
Mozambique Civil	1981-87	350,000	87%	51,000	13%	401,000
Sudan	1984-87	5,000	50%	5,000	50%	10,000
South*	1987-88	250,000	96%	10,000	4%	260,000
	Sub Total	255,000	94%	15,000	6%	270,000
Uganda:Amin	1971-78	300,000	100%	0	0%	300,000
Liberation		---	0%	3,000	100%	3,000
UNLA / NRA	1981-87	100,000	98%	2,000	2%	102,000
	Sub Total	400,000	99%	5,000	1%	405,000
Total		1,720,000	92%	151,000	8%	1,871,000

Compiled from : Sivard R. L. *World Military and Social Expenditures 1987-88*, page 31, *World Priorities*, Washington D.C., 1987
*Source : Press reports for 1987-88

Health facilities and personnel

Medical services in Uganda, the Sudan and Mozambique evolved from the colonial period and are deeply rooted in the western tradition. Medical facilities, for the most part, followed the expansion of colonial armies, administration and commerce and missionaries who built up a significant number of facilities and services. Similarly, government defence forces concentrated in these administrative centres. There is a two-sided threat to the medical facilities as well as to the professionals who provide health services when there is civil war. First, government troops often requisition health facilities if they are in close proximity to either permanent barracks, police posts or temporary encampments. Conversely, if a mission hospital, for instance, is on the outskirts of a district town or located in a rural area, it becomes easy prey for anti-government guerrilla units for either services, provision of supplies, or actual use as an all weather command post. This susceptibility in turn creates suspicion in the minds of the army that such autonomous health units, be they government or mission run, are easily accessible to the rebels and that the health personnel are or may be sympathetic to the rebel cause. Hence medical centres are frequently attacked, looted or arbitrarily closed.

Doctors, nurses and other medical professionals are trained primarily within the western tradition throughout Africa with the advantage that their skills are universally relevant. The disadvantage is that their training makes them readily employable in neighbouring countries or in northern industrialized countries. The brain drain is a problem in normal times and is greatly accelerated in times of civil war or insecurity.

The significance of the impact of war is illustrated in Mozambique where UNICEF estimated that 42 percent of the health centres were destroyed between 1982 and 1986.[6] Virtually all of the health centres in Uganda required rehabilitation as did the hospitals after the liberation war of 1979. Most had to again undergo substantial repairs following the civil war after which the National Resistance Army (NRA) eventually brought stability in 1986. In Uganda, half of the doctors and 80 percent of the pharmacists abandoned their country in search of more rewarding work opportunities between 1972 and 1985.[3]

Mission hospitals did not escape the insecurity. The Catholic hospital at Maracha, north of Arua in the West Nile region of Uganda, closed completely; whereas Kuluva Hospital which belonged to the Church of Uganda lost all its expatriate doctors and only just managed to stay open through the worst atrocities of the war.[7]

The circumstances of war and insecurity drove the German-supported leprosy work to a standstill as well as the Italian and African Medical and Research Foundation (AMREF) Health Training Institute in Wau in the southern Sudan. Voluntary agencies operating in southern Sudan faced the pressure of insecurity which, in cases, prompted a decision to pack up and leave; suspension of funding by a sponsoring agency; kidnapping or threats from anti-government elements or expulsion by the government.[8] Even those NGOs whose purpose it is to provide assistance to children often find that they cannot operate in civil war situations.

Civilians suffer most: Food security systems

Civil war places extraordinary stress on the civilian population. Especially on the most vulnerable - children. Not only are hospitals, clinics, schools and agriculture extension services curtailed or closed but, and perhaps more importantly, commerce, trade and communications are disrupted. When famine struck Bahr el Ghazal province in southern Sudan in 1987, it was caused not just by a localized drought, but because the traditional food security systems were rendered useless due to a whole complex set of issues surrounding the civil war between the government and the Sudan People's Liberation Army (SPLA). While these occurred gradually after the outbreak of hostilities in 1983, there were no food reserves to speak of in much of the south by 1987. Even salt had disappeared, forcing people to use ash as a substitute.

There was little or no food because transport and commerce had ground to a halt. Money was scarce and in limited circulation because the banks and post offices had closed and because the salaried officials such as teachers, para-medicals, local police and others had either fled to larger towns or to the north. What commodities there were, were exorbitantly priced.

In times of hardship such as drought, villagers or pastoralists fall back upon sale or barter of household possessions or livestock, but the conditions of civil war in southern Sudan precluded even this. Livestock were at risk because common and easily controlled diseases took a heavy toll of the herds since veterinary services had come to a complete standstill.[9] Seed, agricultural tools, fertilizers and other inputs were completely depleted and no extension services operated.

Access to the money lenders, who normally provided credit for replanting or restocking was completely out of the question, thus denying credit even at high interest rates. No one from the isolated and insecure rural setting of the south was considered creditworthy under the prevailing circumstances. Seasonal migration for farm or informal jobs in small markets or provincial towns provided the final safety net. But even this posed a grave risk to civilians. There were three armed groups in Bahr el Ghazal in 1987 and any movement of people was immediately suspect. The SPLA wanted the population to remain in the countryside which they controlled. The government army could not tell an SPLA member or sympathizer from the ordinary Dinka, Nuer or Shilluk and therefore tended to treat all equally severely. The militia were the natural rivals to the pastoralist southern tribes affected by the drought who took revenge using their comparative advantage of modern weapons. These militias came from both Bantu tribal groups who lived in the tsetse fly belt of the western area of the south where they engaged in hoe agriculture and the Bagara tribes of Arabs who inhabited the border area to the Bahr el Arab River which marked the boundary between the north and south. The government army provisioned both groups with arms and ammunition to bolster local security and defence against SPLA incursions.

Postal communications as well as the coming and going of family, friends and traders was completely halted. While relatives, government departments, religious groups and NGOs would have provided relief, none of these could visit let alone deliver food or other assistance to the people in Bahr el Ghazal in 1987. The cost to the civilians in southern Sudan was estimated by well placed observers at 250,000 dead; one third of a million refugees; and over a million displaced people to northern Sudan.

Whether in a camp for displaced Ethiopians, Ugandans, Sudanese, Somalis or Mozambicans or in a refugee camp in any of these countries or travelling through a conflict area, it is apparent that food is critical, especially for children who are the first to perish when famine occurs. The whole system of agriculture, livestock herding and trade is disrupted by civil war.

Those left in their country, even in their own homes, are rendered vulnerable when the entire economy goes out of control. This happened in Uganda, Sudan and Mozambique where urban as well as rural people were left without a predictable level of household food security even when they had a fixed income. Inflation gobbled up their salaries. The response was to plant a subsistence garden, even when living in a city such as Kampala. Roadside plots were seen throughout the early to mid-1980s in Kampala and some plots were even cultivated on the edge of the elite Kololo golf-course. The most popular crop in Kampala, Luwero, West Nile, Karamoja or Wau was cassava. Cassava is ideal for such situations because it requires little attention, can be left in the ground and harvested by one person at a time when security permits. Cassava and other root crops are less likely to be looted and are not susceptible to burning in the same way as fields of standing maize or sorghum. Deliberate destruction or looting of crops is common in intense conflict situations. While cassava may be a "war-resistant" crop, it has a nutritional limitation when eaten exclusively. Kwashiorkor among young children increased dramatically in West Nile [10] as did admission of malnutrition cases from the Luwero Triangle and Kampala to the Mulago hospital during the worst war years from 1982 to 1985.[11, 12]

Although we have seen that the traditional food security system did not work for the people of Bahr el Ghazal in 1987, the same is not necessarily true for Uganda as a whole in the early 1980s. Economist Vali Jamal argues convincingly and contrary to Food and Agriculture Organization (FAO) that subsistence agriculture saved Uganda from widespread hunger in the early to mid-1980s, despite massive upheaval in the economy as a whole. His argument is based on national figures and his own personal observations and the perception that a massive shift from cash crops to cassava, sweet potatoes and plantains took place as shown in Figure 3.[13]

Figure 3   **Roots and tuber production in Uganda, 1980-81**

Crop	FAO	National
Cassava	1,583	2,733
Sweet Potatoes	730	1,333
Plantains	3,347	5,900

Source : Vali Jamal, In *Beyond Crisis : Development in Uganda.*
M.I.S.R/African Studies Association, Kampala, 1987

During 1981 to 1986, the child nutritional situation in and around Kampala was monitored. At the various peaks in inflation and escalation of food prices, nutritional surveys were undertaken among pre-school children. None showed malnutrition levels greater than those recorded in the golden years of the 1960s. Ironically, children from poor income groups in Kampala had slightly lower levels of severe malnutrition than comparable slum dwellers in Nairobi. This nutritional observation, while not designed to study or compare nutritional status, nonetheless tends to confirm Jamal's findings and lends support to the view that informal household food security strategies did work in Uganda. There is one notable and important exception: in areas where the population was severely disturbed or displaced such as the Luwero Triangle, West Nile and Karamoja, nutritional status of young children deteriorated in direct proportion to the severity and duration of insecurity.

The following may help explain how the respective traditional food security systems fared in Uganda and the Sudan. Uganda is more densely populated and relies on subsistence and cash crop farming. It has more favourable rainfall patterns than Bahr el Ghazal which is arid, reliant on hoe agriculture and livestock and is less densely settled. Uganda is a much smaller country with major east-west trade routes connecting central Africa to the port of Mombasa. Southern Sudan is largely isolated in its own vastness and lacks any significant infrastructure such as all-weather roads. The bicycle is commonly used by Ugandan traders and householders to transport essential food commodities, whereas the distances in southern Sudan limit the bicycle's use to urban areas and more densely populated regions such as western Equatoria. Finally, the nature and duration of the civil war has had an affect on food availability.

Refugees and displaced

The best known result of civil war is the ensuing plight of refugees. The Ethiopian refugee camps in both the Sudan and Somalia were notorious for the deprivation suffered by their inhabitants and, more recently, the chronic problems of long-term camp life. Similarly, Ugandans who fled various periods of insecurity and armed violent repression became refugees in southern Sudan and Zaire. But refugees have a unique international status which entitled them to relief, sustenance and legal protection under the mandate of the United Nations High Commission for Refugees (UNHCR).

Internally displaced people, on the other hand, have no such rights! Indeed, their fate rests all too often on the government who may not be in a position to provide adequate assistance. Until 1990, there was no international agency, no non-governmental organization and no international law which mandates assistance and protection to the internally displaced, although many NGOs and bilaterals do assist displaced people. It is no wonder then that their condition is too frequently characterized by mass deprivation as reinforced by television reports of the highland camps of war-torn northern Ethiopia in the mid-1980s, the camps of the Luwero Triangle in Uganda and those in southern Kordofan. But the Convention on the Rights of the Child which provides protection for all children offers some promise in providing a frame of reference for reaching this group.

To use just one example, there were 40 distinct displaced camps in and around the three cities of Khartoum, Omdurman and Khartoum North by early 1988. One such community was Hillat Shook, home to 20,000 Nuer and Dinka from the Upper Nile region of southern Sudan. A proud cattle-keeping people accustomed to a rural pastoralist life, they were crowded together in makeshift cardboard hovels built on a garbage dump on the outskirts of Khartoum. No other land was available to them.

Children played amidst broken glass and rusty tin cans. Many families were headed by the mother who scraped out a living in an already overcrowded urban setting with no prospect of regular employment. She had to carry water from a distant public standpipe because the municipal authorities could not provide a public supply. The most destitute women gleaned maize from the lorry park four miles away and brewed illegal beer even at risk of punishment under Islamic *sharia* law.

The health conditions in Hillat Shook and other camps were appalling with low immunization coverage for children, high levels of child malnutrition and almost no preventive or curative services in 1987. While medical services were gradually provided by various NGOs, it is worth comparing the international response to refugees in 1986 and 1987. The programme to assist the 330,000 Sudanese refugees in Ethiopia received funding of $10 million or about $30 for each person. The 1986 budget of UNHCR for the one million Ethiopian refugees in eastern Sudan was $73 million or $73 per person, while the estimated allocation for the displaced in and around Khartoum was only $2.50 per capita.[14] While it is recognized that those may not be comparable in terms of the cost of providing assistance, the per capita availability of funding is meant to be indicative of the neglect of the internally displaced and not to suggest or imply that the needs of refugees are any less or necessarily adequately funded.

The displaced southerners stranded in south Kordofan fared much worse as the news coverage of late 1987 revealed. The horrors of El Meram and Abeyei eventually confirmed that virtually all children under two had died, despite, in the case of El Meram, the presence of a medical team from Médecins Sans Frontières, France.[15, 16]

Since it offered no amenities, the reason displaced southerners inhabited the garbage dump was to be closer to what little employment opportunity Khartoum offered. The government, however, found their presence undesirable and after years of threats, Hillat Shook was cleared in November 1990, its inhabitants ordered onto trucks by the army and dumped 30 kilometres out of town. Hillat Shook was burned to the ground.[17]

Reaching the displaced with medical assistance has proven much more difficult than providing for refugees. For example, a medical team worked in Wau from March 1987 to January 1988 where 50,000 displaced were in critical need of medical services but the medical team had to be evacuated. Difficulties plagued the mission from the outset beginning with basic communication. Initially the radio provided a daily link but gradually the army in Wau became suspicious and demanded that only written messages be transmitted. Later, the radio had to be lodged with the police except for the midday transmission and eventually the radio link became sporadic because of military worries about security. But problems were not one-sided.

Air-freighting medicines and other supplies to Wau was planned on a weekly basis but had to be abandoned after the SPLA shot down a light aircraft (chartered by an NGO providing relief to the south) on take-off from Malakal in neighbouring Upper Nile province. The government responded by closing the air space which left the expatriate medical team stranded without sufficient medicines or food to run the paediatric department in the hospital. The displaced population paid with the lives of their children. Sixty percent of the in-patients died and over 20 percent of the children enroled as out-patients in the feeding centre perished. While the security situation in Wau was particularly bad during the second half of 1987, it was not insecurity which forced the withdrawal of the medical team in January 1988, but the inability to communicate regularly and provision the team.[18] With their withdrawal, an important medical facility was closed, leaving the displaced children with no similar service. Such was the case with many other facilities and relief programmes planned or operated and then closed in Sudan. This was also the fate of similar relief efforts in northern Ethiopia, Uganda and more recently Somalia and Mozambique.

Public health and emergency medicine

In times of civil war, as described above, the collapse of commerce, transport, communications and medical services combine to have an impact on the civilian population. Just as civilians and among them children have become the target in Africa's civil wars, so too have public health facilities and the provision of public health services. In both southern Sudan and Uganda, ordinary hand pumps were damaged or destroyed by government troops during forays into the countryside in search of anti-government guerrillas. Guerrillas similarly destroyed hand pumps in pro-government areas in an attempt to either drive the civilians away or convince them to support their cause. Hospitals, health centres and clinics are also subject to wanton destruction by one or other side.

Likewise, vehicles belonging to the Red Cross, medical teams of various NGOs and even relief aircraft but especially World Food Programme (WFP) food convoys have been the targets of attack. Not only have food and starvation become a weapon but also the use of media by the contestants in civil war. They jockey through the news media to take credit for good deeds and lay blame on their opponents for atrocities which cannot be easily investigated given the circumstances of civil war. When a vehicle carrying a BBC television crew hit a land mine which exploded, killing some of the foreign crew, the southern Sudan became a "no go" area from late 1986 to late 1988. The way in which the media reports on civil war has become part of the agenda and military strategy for both sides in today's civil wars. The significance of relief and media coverage in the quest to reach civilians is so important that James P. Grant, Executive Director of UNICEF, wrote personally to *The Washington Post*, 3 May 1989, in his capacity as Personal Representative of the Secretary-General of the United Nations for Operation Lifeline in an attempt to set the record straight by positively and simultaneously reinforcing the Khartoum government, the SPLA and the Ethiopian government.[19]

Comparative child immunization coverage for northern and southern Sudan graphically displays this inability to provide even the most simple and straightforward of child survival public health interventions (see Figure 4). Three separate localized surveys of IMR in Uganda during times of civil war and widespread insecurity found rates many times higher than normal. In 1980, the IMR was found to be 609 per 1,000 live births in Karamoja [20] while the Luwero Triangle, a Red Cross survey in 1984 recorded an IMR of 305.[21] The IMR in the Kasangati Health Centre catchment area located on the fringe of the Luwero Triangle just to the north of Kampala, was 90 in 1964, fell dramatically to 25 and 33 from 1968 to 1971 but had escalated to over 90 by 1984.[22] While these surveys were small and cannot be implied to be representative of a larger population, they at least represent attempts to collect relevant vital statistics so utterly lacking in civil war settings.

Figure 4 **Comparative immunization coverage in Sudan 1988**

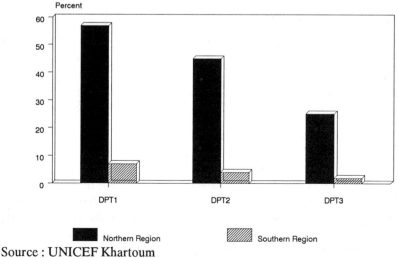

Source : UNICEF Khartoum

Communicable diseases such as measles which can be prevented through immunization rose in both government and mission hospitals in Uganda during the late 1970s and early 1980s. The mission hospital at Kuluva reported a sharp rise in measles admission,[7] while Mulago Teaching Hospital in Kampala recorded measles as the number one cause of paediatric deaths at 29 percent in 1982 and 1983 and 46 percent of post- neonatal deaths.[12] In a review of morbidity and mortality in Ugandan hospitals (both mission and government) in 1981-1982, Alnwick et al.. reported measles as the leading cause of admission in 1981- 1982 compared to third place in 1970 and the second highest case fatality compared to fifth-ranked case fatality in 1970. Measles led the proportion of deaths in 1980-1981 compared to fourth place in 1970 (see Figure 5).[23]

Figure 5 Leading causes of addmission and death in selected government hospitals (in-patients, 1970-71 and 1981-82)

Diseases	Admission		Case fatality rate		Proportions of death	
	%		%		%	
	1981	1970	1981	1970	1981	1970
Measles	12.4	5.3	9.5	4.3	25.6	5.4
Pneumonia	7.3	5.0	8.3	9.4	13.2	11.3
Malaria	9.8	9.1	2.5	3.0	5.2	6.6
Gastroenteritis	7.1	7.2	5.5	5.5	8.4	9.5
Tetanus	0.4	0.5	48.2	46.6	4.5	5.5
Anemia	3.6	3.6	5.4	6.1	4.3	5.2
Dysentery	2.9	0.6	3.5	4.3	2.2	0.6
URTI*	2.7	3.0	3.3	2.7	1.9	2.0
Pertussis	1.4	1.0	7.4	3.4	2.3	0.9
Total	47.6	35.3	6.5	5.5	67.6	47.0
All other causes	52.4	64.7	2.8	3.4	32.4	53.0
	100.0	100.0	4.6	4.2	100.0	100.0

Source : Alnwick *et al.* in *Crisis in Uganda : The Breakdown of Health Services.* Pergamon, Oxford, 1985.
*URTI : Upper Respiratory Tract Infection.

As public health services decline, medical problems increase. As insecurity mounts, fewer doctors, nurses and paramedics are available from the normal health care delivery system. Medical supplies are quickly exhausted and shortages are common-place. Simultaneously, the medical needs of the population increases, sometimes dramatically, if famine conditions set in or if there is widespread fighting, massive dislocation of people or an epidemic. When emergency medical teams are allowed to provide assistance, they all too frequently concentrate on the dramatic life saving interventions such as hospital based services. There is sometimes a lag between hospital-based services and public health interventions such as immunization and restoration of potable water supplies which primarily benefit women and children. Similarly, government and rebel movements alike usually place higher priority on hospital-based medicine than on less-costly public health approaches. However, in the past few years, there has been an increased awareness of immunization as an integral part of emergency medical care. Operation Lifeline in southern Sudan brought immunization services to a higher proportion of young children than ever before.

Note

A comprehensive consideration of health in Uganda, Sudan or Mozambique must take into account the impact of civil war. Civil war contributes to both morbidity and mortality increases and has its worst impact, when combined with natural disasters such as drought, [24] on the most vulnerable and largest single demographic group - children.

An earlier version was published in *Social Science and Medicine*, Vol. 31, No. 6, 1990.

References

1. Sivard R. L., *World Military and Social Expenditures. 1987- 88*. World Priorities, Washington, DC, 1987.
2. Ahmed M., *Within Human Reach: A Future for Africa's Children*. UNICEF, New York, 1985.
3. Dodge, C. P., and Wiebe, P. D.(eds), *Crisis in Uganda: The Breakdown of Health Services*. Pergamon. Oxford, 1985.
4. Dodge, C. P., and Ibrahim, S. A. R., "The civilians suffer most." In *War Wounds* Edited by Towse, N., and Pogrund, B.. Panos, London, 1988.
5. Grant, J. P., *State of World Children Report 1990*. Oxford University Press, Oxford, 1990.
6. UNICEF, *Children on the Front Line*. UNICEF, New York, 1987.
7. Williams, E. H., "The health crisis in Uganda as it affected Kuluva Hospital." In *Crisis in Uganda: The Breakdown of Health Services*. Edited Dodge, C.P., and Wiebe, P.D. Pergamon, Oxford, 1985.
8. Baya, B. K., "Aid at a standstill." In *War Wounds*. Edited Towes, N., and Pogrund, B. Panos, London, 1988.
9. Gonda, S., and Mogga, W., "Loss or the revered cattle." In *War Wounds* Edited by Towse, N., and Pogrund, B.. Panos, London, 1988.
10. Dodge, C. P., "The West Nile emergency." In *Crisis in Uganda: The Breakdown of Health Services*. Edited Dodge, C.P., and Wiebe, P.D. Pergamon, Oxford, 1985.
11. Johnston, A., "The Luwero triangle: emergency operations in Luwero, Mubende and Mpigi Districts." In *Crisis in Uganda: The Breakdown of Health Services*. Edited Dodge, C.P., and Wiebe, P.D. Pergamon, Oxford, 1985.
12. Wotton, K., "Paediatric Mortality in Mulago Hospital, June 1982 to June 1983." In *Crisis in Uganda: The Breakdown of Health Services*. Edited by Dodge, C. P., and Wiebe, P.D., Pergamon, Oxford, 1985.

13. Jamal, V., "Ugandan economic crisis: dimensions and cure." In *Beyond Crisis: Development Issues in Uganda*. Edited Wiebe, P.D., and Dodge, C.P. Makerere Institute of Social Research/African Studies Association, Kampala, 1987.

14. Dodge, C. P., Mohamed, A., and Kuch, P., "*Profile of the displaced in Khartoum.*" *Disasters* Vol. 11, No. 4, Nov. 1987.

15. Dowden, R., "Victims of Sudan's hungry war." *The Independent* 29 Oct., 1988.

16. Perlez, J., "In Sudan, an airlift to a town of misery." *The New York Times* 16 Oct., 1988.

17. Perlez, J., "Sudanese Troops Burn Refugee Camp," "*The New York Times*, 1 Nov 1990.

18. Anonymous, "Sudan's secret slaughter." *Cultural Survival Quarterly*. Vol. 12 No. 2. 1988.

19. Grant, J. P., "Sudan relief effort will save thousands." *The Washington Post* 3 May, 1989.

20. Bielik, R. J., and Henderson, L., "Mortality, nutritional status and diet during the famine in Karamoja, Uganda." *Lancet* Vol. 11, No. 12, Dec., 1981.

21. Dodge, C. P., "Rehabilitation or redefinition of health services." *Social Science and Medicine*. Vol. 22, No. 7, 1986.

22. Namboze, J. M., and Hillman, E. S., "Kasangati health centre: past, present and future." In *Beyond Crisis: Development Issues in Uganda*. Edited Wiebe, P.D., and Dodge, C.P. Makerere Institute of Social Research/African Studies Association, Kampala, 1987.

23. Alnwick, D.J., Stirling, M.R., and Kyeyune, G., "Morbidity and mortality in selected Uganda hospitals.", 1981-82. In *Crisis in Uganda: The Breakdown of Health Services*. Edited Dodge, C.P., and Wiebe, P.D. Pergamon, Oxford, 1985.

24. Dodge, C. P., "African health and medical services." In *Handbooks to the Modern World : Africa* Edited by Moroney, S., Vol. 2. Facts on File, New York, 1989.

CHAPTER 2

WAR EXPERIENCES AND PSYCHOLOGICAL IMPACT ON CHILDREN

Magne Raundalen and Atle Dyregrov

Background

Outwardly, children who survive in a war situation and thereafter receive proper medical care and nutrition seemingly suffer only physical scars. But seen from within the psychological make-up of the child, a different picture emerges. Psychological wounds and trauma suffered in childhood may affect the individual child and, as a consequence, the society for decades. While health and nutritional needs are routinely looked after by relief operations, governments, UN agencies, the Red Cross and NGOs, psychological care is also important.

The first scientific report to confirm this view came from research focusing on long-term after-effects on adult Holocaust victims from German concentration camps.[1,17] More recent research from Finland revealed that a large number have suffered because of psychological problems stemming from their childhood experiences of the Winter War in 1939. Children who were evacuated to Sweden and separated from their families were the most affected as adults. Case studies document that the occurrence of traumatic incidents in childhood can produce anxiety attacks 50 years later.[2,3]

Psychological research provides convincing evidence which is gradually reaching professionals, politicians and the public.[3] For example, the fifth congress of Frelimo in Mozambique included a separate agenda item for education of psychologists and teachers regarding the rehabilitation of children traumatized by war.

Mozambique

The assessment of children's suffering in the little town of Chimoio in the Manica Province of northwest Mozambique was based on questionnaires given to 120 school children, their teachers and many of their guardians. (This project was carried out with the help of the Norwegian Red Cross and resulted in a manual for teachers. Renate Gronvold Bugge played a major role in the data collection in Chimoio.) The children came from four schools on the outskirts of Chimoio and were between 7 and 16 years old; only 31 percent were girls. (We use the terms "guardian" and "caregiver" interchangeably to mean parent or, in the absence of a natural parent, the close relative or person responsible for providing basic amenities.)

Children, teachers and guardians were asked different questions regarding what the children had experienced and how they had reacted. Language difficulties created an initial problem but were gradually overcome through practice with our translator. Questions had to be adapted to the cultural context of Mozambique. While we are confident that our findings are representative of the situation, nonetheless they should be regarded with reasonable caution due to our limited knowledge of Mozambique's traditional society and the difficulties of carrying out this type of research in an insecure setting.

The types of stress the Mozambican children experienced are listed and discussed with the proportion of children reporting the event. The community is as typical as one can find, recognizing that there has been so much internal displacement that almost no community is "typical" in the traditional sense. Chimoio was selected because it was relatively secure during the time we were there. Unfortunately all the causes of war-induced stress mentioned are part of the daily life of many Mozambican children and other children in conflict-ridden areas.[5-16]

Experiencing death of close family members

More than half (54 percent) of the children we interviewed had experienced the death of a close family member. Guardians reported that as many as 69 percent had suffered such a loss. While there is no typical story when the loss of an intimate loved is concerned we heard many accounts such as: "*My brother disappeared. He had gone to war to fight against the bandits. He was shot in his leg, lost much blood and died.*"

One-third (32 percent) of the children claimed that they had seen close relatives being killed. However this was not confirmed by their guardians, as only 18 percent reported that children under their care had witnessed such deaths. Whatever the actual proportion, it means that many children have vivid memories of extremely violent events: "*I witnessed the armed bandits kill my sister's daughter.*"

When we expanded this question and asked how many had witnessed others being killed, we recorded one-third (35 percent) again. However nearly two-thirds (64 percent) had witnessed somebody being badly injured - shot, stabbed or violently wounded. This was basically confirmed by the adults.

The majority of children whom we interviewed in Chimoio had experienced violence and still live with the memories. As much as half of this violence was directed at people they loved and cared for, the rest at people they knew.

Panel 2

"Come on, Mozambique!"

"Mozambique, my home country, is grieving; a grief we didn't ask for. How can we get peace, get rid of racism, apartheid, and all those who try to occupy, conquer, exploit other people's world? I have never seen peace. I don't know what peace is. I have never lived in freedom. But I want a world with human rights. I expect it so I can develop as a person. Even as early as in my mother's stomach I lost my rights. She was forced to work for the colonists in the cotton fields so she had no time to go to the Mother and Baby Clinic for control to treat her future flower. A flower which our former president said should never wither.

"One day in December on Road 1, Avneida de Mozambique, in the Mahala region, my heart almost stopped and the noise almost destroyed my hearing. I met the terrible armed bandits. A military unit stopped and tried to chase the bandits away. Through half opened eyes, I saw cars destroyed, burned bodies all over the place, a terrible and unforgettable event. They destroyed our country.

"I have to ask, don't the Mozambicans have a right to live? We have got a sad life because of this group of terrorists. Their only thought is to violate the human rights; the rights I have got. Why can't we use our knowledge to other things than to eradicate human beings? We need knowledge for development.

"There is now a new law guaranteeing amnesty and forgiving to all the bandits. They can give themselves up to the security forces. And after that, they can live in freedom and together with us to develop the country and make peace plans. But the armed bandits have not approved the proposal. What do they want? In my opinion they have no objective, they are murderers, criminals, thieves. They cause grief and they even attack their own families. Let us get rid of them and after that for sure peace. We are terrorized by a reptile from our neighbour, South Africa. Let's kill it and say, 'Come on Mozambique'.

"Come on I'm only fourteen years!"

Armando, age 14:

Separation and displacement

Because of the war, sixty-two percent of the children had been forced to change their residence. However, only 39 percent had to change school. About one-third of the children were separated from one or both of their parents for more than three months. Half of the guardians confirmed these proportions based upon the children's reports. *"The war has changed my life because I can no longer be with my close relatives, my friends or live where I was born. If it had not been for the war, my family would have been together. We moved because they killed my father."*

Forced and sudden displacement from their place of origin was mentioned very often by children and seemed to be the most common source of stress in the area we studied. More than half of the children (56 percent) reported that their homes had been directly hit, looted or destroyed during the war.

The bandits ("bandit" refers to Renamo anti-government guerrillas also known as Mozambican National Resistance, MNR) have made civilians, especially children and institutions such as schools and health facilities, their primary targets. More than one-third of the children (35 percent) related experiences of their school being directly damaged or destroyed during the war. The adults confirmed these findings. *"Once when I was going to school the bandits came and started shooting. They ruined the whole school. Every time I hear shooting, I start shivering and the shivering continues for an hour every time."*

The terror was not aimed solely at the institutions but included direct violence against the children as well. *"I still have a scar on my left foot from the time I fled. This has changed my life."* Further restrictions meant that the children could not easily or regularly visit family or friends even in nearby villages or cities. Roads were frequently not safe and travel was severely restricted, greatly curtailing the valued inter-personal interaction traditional to Mozambican society.

Kidnapping

Kidnapping was a common occurrence although reliable numbers were very hard to obtain. However, it is believed that as many as 10,000 children from the age of eight upwards were kidnapped by the MNR. Proportionately, this was a relatively small number. In the group we studied, 16 percent claimed to have been kidnapped by the MNR and no one claimed to have been forced to become a child soldier: *"In 1987 the bandits held me prisoner for two weeks. I had to do their dish-washing. They whipped me so badly I still have scars on my body."*

In the study group, 47 percent of the children reported that close relatives had been kidnapped. Upon further questioning it was found that anyone who "goes missing" is assumed to have been kidnapped. This may help to explain why only 21 percent of the guardians confirmed that children under their care had close relatives who were known to have been kidnapped. Because of the interpretation and connotation of this question, we should be careful in drawing conclusions; however, ample evidence exists, that many children have been kidnapped and they certainly have known people who have been kidnapped. *"My mother was taken away for some days, because she refused to give the bandits food. Another time I was kidnapped myself, and taken to different places. I was forced to carry heavy loads, and I was shot at from a helicopter. My life will never be the same."* In a survey conducted in Chimoio in 1989 among teachers, more than two-thirds reported the kidnapping of children from their school districts.

Most of the children experienced repeated shelling and shooting at close range. Around half of the children were caught up in gun battles and had to run or hide. Of these, one in four suffered an injury because of the war. These war wounds resulted in physical handicaps among 10 percent of our sample. *"One day the bandits came and started to shoot. I ran, but was wounded. I have a scar after the wound healed and can only walk by limping."*

Our sample revealed that more than half (56 percent) had personally been threatened by armed bandits and many had a frightening confrontation more than once. Around one-quarter claimed to have been tortured by the MNR. Three out of four children reported that they felt they were going to die during the war. The degree of life threatening circumstances was extremely high.

Looting

"The armed bandits steal different foods, hens and other things. Whoever complains is killed." Looting is common, with material possessions taken, broken or burned; although we have no reliable estimates of how many of our sample actually had their homes looted. We have reservations about responses to this question since relief is often determined based on these surveys and our respondents - both children and adults - may have exaggerated reports of looting.

Discussion

Warfare interferes with children's normal growth and development through the availability of food, housing, health services and schooling, play and family and kinship structures. Parents may not be able to reach medical help in times of illness which results in much higher infant and young child mortality rates, adding to the overall stress of war.

The Mozambique war has had terrible impact on children based upon our research of 120 children in Chimoio. Children living in the cities were usually less affected than those living in the countryside. *"Children in the countryside are always afraid to hear the noises from gunshots. Children in the cities only learn about the war when others tell them about it"* (as one father said in an interview).

The MNR's destabilization approach creates an atmosphere filled with anxiety and uncertainty. Attacks are systematically directed against children and their institutions such as health facilities and schools. Children do not know what will happen next. They fear for the life of their family, for their own life, and for the safety of their village. Fear and insecurity constantly disrupt the structure of the individual, the family and the community as well as the nation.

In the group we studied, 45 percent of the children experienced being without food for more than two days and most of them reported having been through this ordeal more than once. Guardians reported that nearly 70 percent of the children experienced this type of deprivation. *"I do not have food, clothes, shoes, books or other things that the bandits have taken from us."* Some two-thirds (67 percent) of the children reported that they tried not to think and 74 percent tried to avoid associations that made them recall bad experiences. We know from a long history of psychological research that denial may be very helpful right after a traumatic event, but if a child is not given the opportunity to talk through or to work out his or her anxiety, much mental energy is consumed to keep bad thoughts away. This systematic denial may result in long-term problems.

The Sudan case : Essays from Juba

Civil war engulfed Juba from 1983 to the present. While Juba has remained in the control of the Khartoum government, the countryside surrounding Juba is in the hands of the SPLA. Tens of thousands of displaced people have taken refuge in Juba and prices have soared as the SPLA have tightened their stranglehold around the town. However, we managed to get 124 school children between the ages of 15 and 20 to write essays on "Myself in the year 2000."

The pressure of the present was obvious in their writings. The majority wrote: *"If I am not dead"* or, *"If I am alive in year 2000, I will..."* The threat, they felt, was from the war and unrest caused by the tension between the SPLA and the people in the region and the government troops.

They wrote about food shortages and the prospect of famine as well as their concern about the numerous consumer shortages such as schoolbooks, transport, medical care, law and justice, lack of classrooms and unemployment. The impression their essays conveyed was of a "lost generation." Despite dark times, however, many of the students still had hope for the future.

The students clung to the prospect of education and the wish to enroll in the university, not recognizing the weak school system in Juba. Journalists reported from SPLA-controlled southern Sudan that education was a higher priority than even the desire for food and clothing. *"It is an unquenchable spirit that is filling classes as soon as they open in the rebel areas,"* reported Robert Press in an article, "Sudan Goes Back to School," in the *Christian Science Monitor*, on January 26,

1990. The children we interviewed aspired to become ministers, governors, directors and heads of important institutions; some planned to enrol in the Military Academy.

The majority of students from Juba wrote about a personal future with a spouse and children, but some wrote that having a family must wait for the desperate situation of the country to improve. Surprisingly, many wrote with a perspective of three generations in mind: about taking care of their own children, their parents who were getting old (year 2000) and some even pictured themselves as grandparents!

The youngsters stated that if they could live in peace, they would improve the schools and grow enough crops to feed everyone. However, obstacles were presented such as: "*We are bothered by government, or rebels or with other people with guns.*"

God was presented as their strongest ally. Seventy-five percent of the school children placed all their hope in the "*will of God.*" All were striving to be good Christians or good Muslims and thus to be a model for their own future children.

Their parents were prominently thought of in their hierarchy of caring for others. This care was linked to gratitude for what the parents had contributed towards the children's education and what they had forsook to provide for them. The considerable number who aspired to be nurses and doctors also illustrated how high they ranked caring for others.

No one mentioned the government, army, or the liberation forces as a reliable ally for a better future.

These essays represented one-quarter of the students' individual desire to accomplish reforms and changes. The majority of this group wanted to be nurses and doctors and interestingly, several mentioned the sickness of their country. Some specified their intention to build infrastructures such as schools and said that they wanted to instill democracy in people who practised justice to solve the national illness. The level of understanding of some of these students suggested that given the chance, they really could make a contribution towards building a new Sudan.

Interviews with children in Juba

The Norwegian Save the Children Fund helped us conduct 291 interviews with children between the ages of 12 and 16 in Juba. They were interviewed by teachers in March 1989. The interview questionnaire used was based upon the one developed in Kampala in 1986 and used later in Mozambique in 1989. School teachers conducted the interviews in 14 schools located in Juba town.

Twenty-two percent of the children reported the loss of a father; 20 percent, the loss of a mother. Asked if they had experienced the loss of friends and relatives, 89 percent of the children replied "yes."

Events of war

Seventy percent reported shooting in the area of their home whereas fighting near their home was reported by 30 percent. This can be explained, at least in part, by the circumstances of the war. Juba is a garrison town which is controlled by the army from Khartoum. The SPLA, however, controlled the countryside surrounding the town and frequently shelled the city. Therefore, it is reasonable that 70 percent of the children said they had heard shooting in their area but had not experienced fighting at or near their home.

It is clear from the essays and interviews that many of the children in Juba were not only fully aware but had direct experience of the most violent aspects of war - the death of a parent or other close family members. The war was so intense in and around Juba that access was severely limited and the town was frequently cut off for long periods of time. The only regular access from Khartoum or Nairobi was by aircraft. River and road links were entirely severed.

In comparing the two different approaches, essays and questionnaires, we found that the essays were more insightful. While we were not able to carry out an extensive analysis of the Juba data, we have provided a comparative table on the experiences of war and violence between the three countries in the concluding section of this chapter.

Summary of war: Uganda, the Sudan and Mozambique

Figure 6 summarizes data from the three countries based upon a similar questionnaire. Given the difficulties of carrying out research in war situations and the inability to double check the children's reports (other than in Mozambique), these findings should be treated only as indicative. Nonetheless, they should convince everyone that the children surveyed had experienced a lot of war and violence in their young lives.

Psychological impact

The psychological effects of war on children is described in terms of known psychological reaction patterns ranging from aggression and revenge (an aspect which we think is exaggerated) to anxiety, fear, grief and depression. Since these reaction patterns are described thoroughly in our first book, *War, Violence and Children in Uganda*, we will focus here on a less familiar aspect, the Post Traumatic Stress Disorders (PTSD) in children. In medicine, trauma can be described as the acute damage to the body organs or system. When we use trauma to describe the psychological impact of events, it is more difficult to define, and there is not the same general agreement on what psychological trauma is.

Figure 6   **Experience of war questionnaire**

Experience of War Questionnaire administered by teachers to children 10-16 years old	Juba Sudan 1989	Kampala Uganda 1986	Chimoio Mozambique 1989
Number sampled	291	297	120
Loss	Yes	Yes	Yes
Death of			
Father	22%	10%	13%
Mother	20%	4%	9%
Other close relative	85%	51%	54%
EXPERIENCE OF VIOLENCE			
Gun battle near home	30%	64%	67%
Personally shot at	19%	7%	27%
Own or neighbour's house looted	23%	61%	56%
Known neighborhood rape victim	32%	61%	N/A
REACTIONS ON WAR			
Are you afraid of war	99%	83%	75%
Have you had to hide	56%	66%	56%
Are you anxious every day	66%	46%	56%

When a child witnesses his or her mother being killed, everyone easily accepts that this is traumatic. But less gruesome events can also be traumatic for children. If soldiers threaten to kill a child's mother and the child believes that they will do so, this is also trauma. When a psychological trauma is defined, it is important to remember that it is the child's subjective experiences, the meaning that the child ascribes or gives to the event that is important. Sometimes a situation which we think the child may have forgotten is still painful.

A traumatic event is often associated with overwhelming affects so children in adulthood will try to avoid events associated with the affect. It hurts to be reminded of what happened, and repression takes place to avoid painful feelings.

Originally, in child psychology, the understanding of childhood trauma centred on separation anxiety; the young child's feelings of being abandoned or left alone, and parental inability to relieve the child's distress. In psychological literature, a trauma refers to intense, sudden events that overwhelm the child's capacity to cope with the memories and feelings that are triggered.

Traumatic event

Psychic Trauma is when an individual is exposed to one or a series of events that cannot be assimilated or integrated in the child's basic assumptions of the world. It is the subjective experience of the event, more than the objective event that determines which event(s) this will be. The child gradually develops basic assumptions about, for example, how safe the world is; about justice and/or injustice; whether people are trustworthy or not, kind or unkind; about one's own worthiness and/or unworthiness; about courage and fear. These assumptions may be shattered after experiencing trauma.

Most studies and theories concerning trauma have focused on trauma as a single event. We believe children that suffer repeated losses or extreme stress due to situations such as kidnapping, harassment and force, will experience cumulative or serial traumas. Although children are able to live through difficult situations without breaking down, there is no reason to believe that children can escape the harmful effects of traumatic situations. Repeated exposure to such situations will probably cause more permanent damage to their personality formation, than single events. This is why it is so important to be able to help children that are exposed to the traumas of the magnitude experienced in Mozambique, the Sudan and Uganda and other war-ridden countries.

Post traumatic stress in children

Post Traumatic Stress Disorder or PTSD, in the classification system DSM III R, can readily be identified in the following way:

The person has experienced an event that is outside the range of usual human experience. Such an event would be markedly distressing to almost anyone. Examples include: a serious threat to one's life or physical integrity; serious threat or harm to one's children, spouse, or other close relatives and friends; sudden destruction of one's home or community; witnessing someone being seriously injured or killed as the result of an accident or physical violence.

The traumatic event is experienced over and over again in at least one of the following ways: recurrent and intrusive distressing recollections of the event (in young children, repetitive play in which themes or aspects of the trauma are expressed); recurrent distressing dreams of the event; sudden acting or feeling as if the traumatic event were recurring (includes a sense of reliving the experience, illusions, hallucinations, and dissociative--flashback--episodes, even those that occur upon awakening or when intoxicated); intense psychological distress at exposure to events that symbolize or resemble an aspect of the traumatic event, including anniversaries of the trauma.

Persistent avoidance of stimuli associated with the trauma or numbing of reactions in general (not present before the trauma), as indicated by at least three of the following: (1) Efforts to avoid thoughts or feelings associated with the trauma. (2) Efforts to avoid activities or situations that arouse recollections of the trauma. (3) Inability to recall an important aspect of the trauma. (4) Markedly diminished interest in significant activities (in young children, loss of recently acquired developmental skills such as toilet training or language skills). (5) Feeling of detachment or estrangement from others. (6) Restricted range of affect, e.g., unable to have loving feelings. (7) Sense of a foreshortened future, e.g., child does not expect to have a career, marriage, or children, or a long life.

Persistent symptoms of increased arousal (not present before the trauma). Arousal means that bodily systems are active, like the body is prepared to fight or run away. This is indicated by at least two of the following: (1) Difficulty falling or staying asleep. (2) Irritability or outbursts of anger. (3) Difficulty concentrating. (4) Hyper-vigilance (always prepared or on the look-out for danger). (5) Exaggerated startle response. (6) The body reacts when an event symbolizes or resembles an aspect of the traumatic event, e.g., a child whose family was killed in a fire reacts with intense fear and anxiety (cry, scream or become paralysed) when there is smoke or signs of a fire.

Some other characteristics known to be more common among children than adults include the following:

- The child re-enacts or performs acts similar to what they did during the traumatic event;
- Something that occurred after the event can be misperceived to have happened before the event, an event can be thought to have lasted longer than it did;
- develops more long-lasting personality changes;
- usually does not show stress-induced amnesia (i.e. children do not forget what they have experienced);
- expresses little disbelief about the reality of events experienced or witnessed;
- children may try to reverse traumatic helplessness and anxiety through fantasy and rationalization (as we learn more about children's post-traumatic reactions, the definition of PTSD be made more specific for children).[18-20]

Separation, displacement and killing

Forced separation from parents or guardians and displacement from home cause traumatic stress reactions. Sometimes years go by before children and families are reunited. In Uganda, we found that the children who witnessed the killing of close family members, not only experienced grief, but also specific traumatic after-affects. Other causes of stress such as bombardment, shelling or firing at close range also result in post-traumatic reactions in children. After the Second World War, the most common reaction was anxiety to stimuli associated with warfare, such as sirens.

Reactions to trauma

Some of the most common reactions children have following exposure to extreme stress are listed below:

- fear and anxiety
- intrusive images and thought
- difficulties concentrating
- isolation
- sadness and depression
- avoidance behaviour
- guilt
- traumatic play and re-enactment
- sleep disturbances and bad dreams
- physical symptoms

Conditioned fear

Children develop fear and anxiety following traumatic events which may be expressed as:

- conditioned fear
- fear for parents and guardians
- fear of separation
- more general fear and anxiety (of being alone, of the dark)
- hypersensitivity

The most common form of fear is conditioned fear; when children react strongly to all stimuli that directly or indirectly remind them of the event. For example, children who have been kidnapped may react with intense anxiety when they see someone who resembles their original kidnapper(s). If they were kept in a special hut or tent, had to eat a particular food, or had to perform certain acts, any of these associations could lead to intense fear. These fears may be part of horrifying nightmares, and result in the children re-experiencing parts of the original trauma.

Things not directly associated with the event may also trigger reactions, such as, if the child has experienced gunshots, loud noises can trigger intense fear, and the readiness to act.

Following the death of a sibling, a parent or guardian, a child's anxieties are very much centred on the remaining parent or guardian. Children become anxious when separated from their parent or guardian, or when going to sleep. In the group we studied in Mozambique, 73 percent of the children were frightened that something would happen to their loved ones, and 76 percent were afraid of being separated from their family.

Children may fear being alone, they may be afraid of open spaces, and some experience acute anxiety and scream or call for help. One boy said that the worst experience he had been through was the bombing of the area where he lived. During the attack, his father took him and his siblings and hid under a tree. Since then, he cries a lot.

Hypersensitivity is the condition when children are watchful and alert, easily upset or startled by sudden noises or changes in the environment which tend to be interpreted as danger signals. Almost half of the children we studied, reported that they became afraid without reason, and around 70 percent said that they worried a lot. *"War is terrible. I am afraid of everything because of the attacks that have happened close to us."*

As parents or guardians become fearful, so do children and their reactions are a response to their parent's behaviour. When parents expect the worst to happen, so will children. Fear is not all bad and may help a child to stay away from dangerous situations. However, if the fear is out of proportion to the situation, it can have negative consequences for the child and interfere with concentration and social contact.

Children experience intrusive recollections of traumatic events. This means that parts of their traumatic experiences keep surfacing into their consciousness, without the child having control over these memories. These intrusive memories are especially bothersome at night when activity slows down. Sixty percent of the interviewed children, had dreams about events they had experienced. The memories and the anxiety interfered with sleep, and with the ability to concentrate at school. *"One day the bandits came and they killed my neighbours, and this has highly upset me. After they started to terrorize us, I sleep worse than before."*

After being involved in or witnessing a traumatic event such memories are common, especially in older children, but also in children under two years of age. Children often keep such memories to themselves and do not talk with adults about them until directly asked. Although these memories become less frequent over time, for many children they remain unless they have the opportunity to talk or play them through.

The killing of close family members leaves highly accurate and detailed visual images. In the group we studied from Mozambique, 78 percent of the children reported that when they saw or heard something that reminded them of the event, they became very upset; 67 percent felt as if the event was happening all over again.

Given such painful memories, it is not difficult to understand why many children show a decline in their school performance or their work at home. Some children stay at home to make sure nothing happens to their parents. Many teachers are unaware of the effect that trauma has on children's moods and on their ability to concentrate in school.

More than 60 percent of the children we studied reported that they were reluctant to play with other children. While a reduced interest in the outer world can be a sign of depression, it may also be a sign that the child is afraid to be outside because of what may happen. Some children also distance themselves from their parents or friends in order to be alone with their feelings. Sadness and crying are common reactions. In Mozambique, 61 percent of the children said that they cried easily; girls cried more frequently than boys and 55 percent of the group said that they sometimes or frequently felt sad or depressed.

Different types of avoidance reactions are sometimes part of children's reactions to trauma, such as staying away from all situations that remind them of a traumatic event, or demanding that adults not talk about the event. Sometimes children become phobic. Fear generalizes many activities and situations that only slightly resemble the original fear-provoking situation. The children cannot participate in any behavioural activity that is associated with the traumatic event. For example, they cannot be in the place where the event happened; are unable to hold or see a gun or anything resembling guns.

In Mozambique, 67 percent of the group we studied reported that they tried not to think about events they had experienced, and 74 percent, tried to avoid reminders of such events. Denial may be helpful immediately after a traumatic event, but if a child is not given the opportunity to talk and has to use a lot of energy to repress traumatic memories, this may result in long-term problems.

Guilt

Children may feel guilt because they failed to provide help, and remained safe when others were hurt, or because they were unable to protect siblings or parents. If children are forced to commit bad acts against their own community or family, this can also cause guilt. Although they may have known that they would be killed for failure to follow an order, they may nonetheless blame themselves for what they were forced to do. These ambivalent and mixed feelings are difficult to handle and children sometimes cope by shutting off their feelings. Cultural beliefs, such as belief in one's ancestor's protection, may help in such situations but it can also cause extra confusion. Usually the presence of guilt will increase the severity of other reactions. Particular care should be taken to help children who feel guilty.

Play

Younger children will repeatedly deal with traumatic events in their play or drawings. They may make toys clearly related to the event, or act out parts or the entire traumatic event in individual and collective play. This, of course, is a way of mastering both the cognitive and emotional aspects of what they have experienced. Play can be very useful in helping the child work through bad events.

While play usually can give children some relief from feelings, especially anxiety, they are unable to get any relief from anxiety when the play is repeated again and again in the same manner. Such compulsive play tends to increase the anxiety, and adults need to help the child to alter the pattern of play in a way that will help the child gain some mastery of the situation. This may be done by joining in the child's play and altering the sequence of events, helping the child, for example, to give a different ending to the event.[18]

Often children identify with the forces that control the dramatic events. Through play, they are able to handle fear and anxiety in their fantasy and thus change the course of events and have control over what is happening. At times they take the role of those in charge, of those in power. Although play is favoured by younger children, it can also be seen in adolescents, who, through written expression, may seek to master the trauma.

Play can also involve drawing. Through drawings, children can participate in wish fulfillment, express feelings that are difficult to put into words, and are able to deal with a traumatic event in a symbolic manner. Role play and singing also help represent what they have experienced.

Re-enactment of a traumatic situation is more than just dealing with the trauma in play. The traumatized child can include the same physical activity, attitudes or fears that took place right before or during the trauma. The child may also show the same physical reactions during the actual event (for example fainting, daytime wetting, cramps), and they may re-enact things that took place immediately before the event (arguing with friends or adults). Such reactions indicate that the child needs active help.

Dreams

Children may be reluctant to go to sleep for fear of dreams or nightmares. They may demand company until they fall asleep, they may want to sleep together with others, or desire to have a light or fire. Even adolescents can react in this way.

Most children will have dreams related to what they experienced following traumatic events. Parents and guardians describe this as common among children who have been kidnapped. Some children have terror dreams which they can not recall once they are awake. Children can also other dreams where something terrible happens or is about to happen, including dreams involving their own death. Traumatic dreams bring little relief, but when they are talked about and the child is given some understanding of why such dreams occur, this can help in preventing them. As children are able to put their feelings into words, night terrors tend to disappear.

Children also experience physical symptoms following traumatic events such as headaches, stomach aches, muscle tension, and body aches, especially in the neck and shoulder. Younger children can regress and start bed wetting again. In addition, eating habits and weight may change after traumatic events.

In Mozambique, three-quarters of the group of children we studied said they felt better after relating their experiences. However, 28 percent of the children reported that they had never talked with adults about their experiences, and another 40 percent had only rarely talked. One-third had frequently talked with adults about their experiences. It is interesting to see that as many as 83 percent of the children reported that they had talked with friends about their experiences and had talked more with friends than with adults.

References

1. Rosenheck, R., "Impact of Post Traumatic Stress Disorders of World War II on the Next Generation," *Journal of Nervous and Mental Diseases*, 174(6), 319-327, 1986.
2. Lagnebro, L., *Finnish Children of War*, Institute of Migration, Turkul, Finland (in press), 1991.
3. Boothby, N., "Children in emergencies." In Ressler, M.E., Boothby, N., Steinbock, D.J., *Unaccompanied children in emergencies. care and placement in wars, natural disasters and refugee movements*, New York, Oxford University Press, 1988.
4. Palme, L., "Why the Childrens' Convention?" Speech given at a UNICEF and Swedish Save The Children conference on The Convention on the Rights of the Child, October, 1988.
5. Punamaki, R.L., "Childhood in the Shadow of War. A Psychological Study on Attitudes and Emotional Life of Israeli and Palestinian Children." *Current Research on Peace and Violence*, 5, 26-41, 1982.
6. Raundalen, M., Dyregrov, A., Mugisha, C., Lwanga, J., "Four Investigations on Stress Among Children in Uganda." In Dodge, C.P., Raundalen, M. (Eds.), *War, Violence and Children in Uganda*. Oslo. Norwegian University Press, 1987.
7. Bilu, Y., "The Other Nightmare: The Israeli-Arab Encounter as Reflected in Children's Dreams in Israel and the West Bank." *Political Psychology*, 10 (3), 1989.
8. Bryce, J.W., *Cries of Children of Lebanon: As Voiced by their Mothers*, UNICEF Regional Office, Amman, 1986.
9. Caines, E. and Wilson, R., "Coping with Political Violence in Northern Ireland." *Social Science and Medicine*. 28 (6), 621-624, 1989.
10. Comite de Defersa Los Derechos del Pueblo, C., "The Effects of Torture and Political Repression in a Sample of Chilean Families." *Social Science and Medicine*. 28 (7), 741-740, 1989.
11. Dawes, A., "The Effects of Political Violence on Children: A Consideration of South African and Related Studies." *International Journal of Psychology*. 25 (1), 13-31, 1990.
12. Gustafsson, L., Lindkvist, A., *Krigens Barn (Children of War)*. Kommuneforlaget, Oslo, 1989.
13. Milgram, R., Milgram, N.A., "The Effects of the Yom Kippur War on Anxiety Level in Israeli Children." *The Journal of Psychology*, 94, 107-113, 1976.
14. Punamaki, R.J, and Suleiman, R, "Predictors and Effectiveness of Coping with Political Violence Among Palestinian Children." *British Journal of Social Psychiatry*. 29(1), 67-77, 1990.
15. Rosenblatt, R., *Children of War*, Doubleday. New York, 1983.

16. Rohnstrom, A., "Children in Central-America: Victims of War." *Child Welfare,* 68(2), 145-153, 1989.

17. Dasberg, H., "Psychological Distress of Holocaust Survivors and Offspring in Israel, Forty Years Later: A Review." *Israel Journal of Psychiatry and Related Sciences.* 24 (4), 243-256, 1987.

18. Pynoos, R.S., Eth, S., "Witnessing Violence: Special Intervention with Children." In Lystad, M.,(ed) *Violence and the Family,* Brunner/Mazel, New York, 1986.

19. Terr, L., "Children of Chowchilla Revisited: The Effects of Psychic Trauma Four Years After a School Bus Kidnapping." *American Journal of Psychiatry, 140,* 1543-1550, 1983.

20. Eth, S., Long-term Effects of Terrorism on Children. *The Western Journal of Medicine.* 147 (1), 1987.

CHAPTER 3

FLEEING THE WAR
STREET BOYS IN KHARTOUM AND MAPUTO

Cole P. Dodge and Magne Raundalen

About 10,000 children are living rough, sleeping in the dust of Khartoum's streets. They are called *shamassa,* the tattered vanguard of the displaced southerners forced to leave their rural homes after years of civil war in the south. Sudan's street children are almost all boys, and, although some of them talk of returning to their villages, for most, the street is their home. The only home they will have for the rest of their childhood. They survive by begging, stealing and working legitimately but for long hours and low wages. They are children with guts and courage who dream of a better future for themselves and their fellows.

As food became scarce, older sons - although they may have been only nine or ten years of age - were sent to Khartoum to fend for themselves. The boys, encouraged by stories of the exciting big city, rode on trucks or on the roof of trains to the capital. When they arrived, they found few jobs available and their only source of food was a garbage heap. In desperation, they begin to beg and steal and some to prostitute themselves for the equivalent of 50 cents.

Most of the boys have fled the south because of the civil war, fearing for their lives after their brothers or fathers were killed, forced to flee or joined the SPLA. Their hope is to go back home, but that is possible only if peace returns. Khartoum street children have kept their good nature and resilience in the face of great odds. The terrible situation make it necessary to steal, but perhaps the difference between survival and criminality needs to be drawn.

The war has not only driven young boys from the south but their mothers and fathers as well. In 1987, a survey was conducted of 23 settlements where displaced people live in Khartoum.[1] The results reveal that 25 percent of the inhabitants are displaced by drought and desertification, while the rest were driven from their homes by civil war. These displaced people fabricate huts from discarded cardboard cartons while one settlement was built on the garbage dump because all other locations did not provide security of tenure (see chapter 5). The rate of malnutrition among preschool children is more than double the national average. Children from these communities also go into the streets in search of money to help support their destitute parents, thus competing with the street children who have no home or shelter in the city.

While the number of street children varied from month to month, the impression of social welfare agencies working exclusively with them was that about two-thirds were from the South and one-third from the drought-affected northern regions which is quite similar to the one-quarter northern and three-quarters southern found among displaced families in the survey. A major difference between the situation of street children from the north and south is that the northern boys had the option to return home safely, while all of the south was unsafe due to civil war and inter-tribal feuding left no place for the southern boys to go.

Prior to 1984, street children were virtually unknown in Khartoum, and were limited to a few shoe-shine and market boys who came from nearby slum families. However, estimates range from a low of 5,000 up to 20,000 boys who flooded the capital in mid 1980s. By 1990, street boys were a predictable part of the urban landscape. Due to their visibility, the international press has reported on them periodically since the mid 1980s but never with an in-depth appreciation of the life long impact their life on the street will mean to them or their society.[2]

Street children are almost exclusively boys between six and 16. They come to Khartoum in search of money, but only find more hardship, as well as vice and crime. There is very little of the golden streets, jobs and easy life they had dreamed. Rather, the roads are pot-holed, dusty and littered with uncollected urban refuse. Drains are clogged and smelly and even water is scarce and only sporadically flows from public taps. Arriving children are thrust into an already crowded urban environment and must engage in menial tasks such as carrying purchases for shoppers in the open markets, washing cars, hawking cigarettes, begging or, stealing.

The few street boys who manage to find jobs must work long hours for low pay. Market boys carry baskets for the tips given by shoppers and merchants, and have to compete with displaced and otherwise unemployed men. In the market, the street boys have no rights and are consequently subject to exploitation, such as unloading a truck for a vegetable wholesaler and then given only a pittance. They also stand the risk of arrest by the police who periodically sweep the market looking for vagrants. Vagrancy is illegal in Sudan. A payment of about two dollars buys an immediate release, but boys with money in their pocket also risk losing it all. Boys without money are carted off to prison or the reformatory. If no friend or relative comes to rescue them, they can spend days or even weeks in detention.

Street children quickly become street wise and organize themselves. Most boys live in groups which are reasonably well organized. Older boys are the leaders and form a tightly woven unit of between four and ten younger boys. Not only do leaders demand obedience but also a share of their daily earnings. The younger boys, especially six and seven years old, are detailed to begging while older boys stake out a section of the curb and collect tips for watching or washing cars. Older boys may engage in petty theft, such as nicking easily removed wind-shield wipers, tail lights or mirrors from parked cars. By mid-afternoon, the older wiser street

Panel 3

"Jaksa is a special boy. He came, one evening, to my ice cream shop, Casa Blanca, and just stood there watching. I sort of yelled at him, for there were many boys, and what is the matter with you now?'

"Very quietly, his eyes downcast, he lifted a filthy trouser leg and I felt like fainting. His leg was a horrible mess of pus. My first reaction was to sit down and cry. However, I took him in my car and drove him to Dr. Ali's clinic. I opened all the windows and put on the air conditioner but still felt sick from the horrible smell of rotting flesh. Dr. Ali's first thought that the leg might be cancerous and he wrote a letter for me to take to the head of the skin diseases department at the Khartoum General Hospital.

"I left Jaksa in front of the hospital and told him to wait for me there. The next morning, I took him to see the doctor. After one look, I was told to take him away. His leg smelled so badly that he didn't want the boy near him. My short temper exploded and I told him what I thought of his kind. Back to Dr. Ali, God bless him, Jaksa was put on the table and Dr. Ali's assistant held his hands by lying across the table and I held his head while Dr. Ali sliced away at the rotting flesh. In order to find where the rotten flesh ended and healthy tissue started, the doctor could not give Jaksa any anesthetic. Jaksa screamed and I screamed, perhaps, even louder; yelling at Dr. Ali, yelling at Jaksa, yelling at the world that allows children to endure such suffering!

"With treatment of heavy doses of antibiotics, Jaksa's leg got well. By the time I left for my annual leave, Jaksa only needed skin transplants and was admitted to Omdurman Hospital. There was not enough food there so Jaksa left before the transplant could be done. By the time I came back, Jaksa's leg was almost as bad as it was before. Again Dr. Ali treated it, but it did not respond and in the end, Dr. Ali said the leg would have to be amputated to save the boy's life.

"Jaksa was reluctant to agree to the amputation. A few days before the operation was due, he asked me to try and get something for him - a pair of shoes. He had never owned a pair of shoes and before his leg was cut off, he wanted a pair of shoes. I bought him a pair of shoes. On the day Jaksa was to have his leg amputated, Dr. Ali again examined him carefully and decided not to cut off the leg but to try and cure it again. The leg got better. Jaksa (the name he called himself after the best known football player in Sudan), attended school, learned to be a carpenter and now plays football every free minute he has."

From Blanka el Khalifa's recollections, Khartoum, Sudan, 1988.

boys begin their trade of selling glue and petrol. These dealers move amongst the street children selling a "cap" of glue or a "rag" of petrol, first to the leaders, who in turn share out a small portion to their group. A productive member may get an extra portion and is then able to "numb" out the reality which engulfs him.

Escape to Maputo

The situation for children in Mozambique has gradually deteriorated through the years. Mozambique obtained its independence in 1975 after Front for the Liberation of Mozambique (Frelimo) forced the Portuguese colonial power to leave the country. The liberation of Mozambique was considered a threat to Rhodesia, now Zimbabwe and Ian Smith's regime armed and trained a destabilization force (later called Mozambique National Resistance - MNR or Renamo) to fight Frelimo troops. With the emergence of an independent Zimbabwe in 1979, the Renamo lost much of its support from the local population and was on the point of disbanding when South Africa intervened.

For a decade, the armed bandits, as they are known in Mozambique, have continued to destroy and destabilize the country. In many provinces, more than half of the schools and health clinics have been bombed or destroyed. At the end of the decade, almost one million Mozambicans lived in refugee camps in Malawi and the figures of internally displaced people are estimated to be even higher. Although the South African government had officially declared the cessation of their support for Renamo, supplies were still being received.[3]

Renamo has disrupted the transport system and thus the delivery of relief food. Thousands of families have no possibilities of returning home because of continuing insecurity. Most provinces can only be reached by plane. Air transport is far beyond the reach of the ordinary people, especially displaced children.

The situation in the cities has also deteriorated for street children with the continuous decline in the country's economy. According to UNICEF Mozambique, the situation for urban families deteriorated dramatically following the Structural Adjustment Programme imposed by the International Monetary Fund (IMF) which the government had to accept.[4] The noticeable result for the street children was less money and shortly afterwards a decline of goodwill from the pedestrian on the street. Now the street children had to count on the sympathy of the international residents in Maputo.

This shift of fortune resulted in, from necessity, more aggressive and antisocial behaviour patterns among the street children. They were forced to steal and fight for their survival, and therefore came in conflict with adults and the police more frequently. This was very different from the former helpful and happy go lucky street child who the public had grown accustomed to. A wave of dislike for the street boys swept Maputo and they were pushed outside the city. It was not difficult to link macro-level events represented by the new economic policies directly to the micro-level life of children living on the streets of Maputo. This policy created a new type of street child who came from the urban slums, was already marginalized but now became desperately poor. Often, stronger street boys rob the younger ones, taking their clothes during cold winter nights.

The street children of Maputo are all boys which is what we also observed in Khartoum. Why? When we asked the street boys, they stated that girls were also displaced, however, they were not allowed on the streets but placed in homes to work as servants. They were severely punished if they moved on the streets. Many of the street boys concluded that they were lucky compared to many of the city-girls they knew. Girls were placed by parents in the care of relatives or in better-off residences as housemaids.

Methods and perspectives for investigation

Do we know why children become inhabitants of the streets of the big cities in Africa? In the case of Khartoum the problem of street boys was almost non-existent before the extensive drought in the west and the escalation of the war in the South. Therefore the answer to the recent influx of children to Khartoum was simply drought and war.[5,6]

In Mozambique, the street children arrived when drought and hunger occurred in combination with terrorism, atrocities and war perpetuated by the MNR bandits. Again the simple answer to the street children issue was: "In Maputo it is caused by macro-events such as war and drought."

In our surveys of street children we have gone a step further than just identifying the reason why and from where the children came. When we talked to the children, listened to their wishes, fantasies and prospects for the future and heard what they would do if they had a son who ran away from home, we gained further insights. Fundamental factors re-emerged in combination with related but more subtle changes such as female-headed households.

In Sudan, the influx of street children could not be seen in isolation from adult male migration and very recent female migration which provided a strong model for their children.[7,8] A parallel existed in Mozambique caused by the considerable number of men who were guest workers in Zimbabwe, South Africa and Swaziland.

Surveys in Khartoum indicated that most of the northern boys had come from Nyala. Yet a visit to Nyala, in western Sudan, made it clear that only a very small percentage of the age group had left. The same held true for the southern region of the Sudan. Only a fraction of the southern boys became street children.

Another general assumption among social workers who worked with street children and a strongly held opinion of the general public was that street children came from the poorest families and that they were illiterate, lazy, school dropouts who were intrinsically untrustworthy and anti-social in their behaviour. In fact, our findings based on interviews with their parents, neighbours and teachers were quite contrary to these general conclusions. Many of the street children came from rather resourceful backgrounds and undertook to either follow in the footsteps of their fathers or to manage on their own with considerable ingenuity. We found that the street boys from the south did not necessarily come from the most war-affected villages. Why, we ask ourselves?

Push and pull

The explanation of the influx of children to the big cities is usually described in terms of what we call "push-factors" such as famine, caused by drought and desertification, or war and insecurity, all of which contribute to a breakdown of the family. However, we found a different set of reasons which we referred to as "pull-factors." These were drawn from projective interview methods wherein we invited the children to share their fantasies about a cartoon character who resembled street boys in Khartoum. They told us of the freedom they imagined on the streets, a romantic image of urban life derived from the movies.

Other boys who had seen and described the big city provided a powerful reason for the desire to leave home. However, the children also thought of the positive opportunities that the city offered such as schooling, jobs, money and marriage, not only the negative explanation so quickly provided by aid workers, government officials and social welfare agency personnel.

Our approach and methods for studying the psychological forces at work among street children had to include an appreciation of the break-up of families, changes in the village community and sometimes the national perspective.[9,11]

First we used the traditional method of interviewing the children concerning family, background, life at home, condition on arrival, time and life in the street and contact with family. For insight on the psychological forces, we constructed assessment techniques for exploratory use since we had no previous culture-specific guideline to proceed from.

Five cartoon drawings with typical Sudanese and Mozambican scenes were shown to the boys which covered a series of topics from their home village to the journey to the city. The children were asked to tell a story about a boy of about their own age, and asked what he thinks, how he feels and what happens to him. The last picture showed two boys sleeping side by side in the street. They are prompted with a question such as: *"Yameen sleeps in the street and he is dreaming, what is he dreaming about?"*

Culturally-specific hypothetical questions which provided insight into future prospects were also asked. *"If God (Mozambique) or a "genie" (Sudan) suddenly stood in front of you and gave you three wishes, please tell me what your first would be and why?"* Another question was used, depending on how willing a particular child was to the interactive process: *"Can you tell me about a dream you had, a dream you remember?"*

Both in the interview and in the projective questions we tried to solicit answers which revealed the children's view of their future. *"What do you want to be in the future?"* *"What about your future, the future of the nation and the future of the whole world?"* *"If you were a father and your son ran away to the street, what would you do then?"*[10]

Why do the Sudanese children leave home?

The majority of those interviewed told about trouble in the family--a parent's death, the father leaving home or remarrying. Although many mentioned famine, war, drought and poverty when talking, less than a third attributed running away from home directly to macro-events. The "push-factors" dominating their answers to questions, were contrasted to the stories they told about the cartoon sketches. These latter responses followed three recurrent trends: First the boys talked about "pull-factors" such as searching for a school, finding a job and getting more to eat. Secondly, they described a positive and good friendship with other boys living in the streets. And, third, when referring to the drawing where the boy returned home, they suggested that the boy recollected street life and thought about returning to his friends. Again the pull-back theme: in some ways they wanted to return home, but they also had a realistic enough understanding of their home situations not to overstate the welcome they would receive.

The boys who came from southern Sudan[11] reported witnessing people being killed and family members lost, disruption of the school system and loss of educational opportunities. Although some of the children also referred to remarriage and family conflicts and rejection, more often they cited the war as their primary reason for leaving home. Some of them even mentioned that they collaborated with their parents to get away: *"My mother put me on a truck heading for Khartoum and said 'I want to think that one of my children survived this war!'"*

Discussion

The dominant theme was the break-up of the family and inadequate household resources. Personal reasons revolved around psychological self-esteem, friendship and prospects for the future in the street.

Interviews with street children, school teachers and social workers about the possibilities after years of life on the street yielded some remarkable insights. All of the children related to "instant resocialization," that is immediate rejection of such street behaviour as stealing, fighting and glue sniffing, once they were put into a new and desirable situation. Asked why, the street children stated that school was what they came for, and now they had it, they understood the proper behaviour. *"You have to steal food to survive, but you cannot steal a school."*

Many social workers stated, however, that the limit seemed to be about four years on the streets--after that the boys were less prone to respond positively to resocialization, reunification or schooling programmes.

To understand the phenomena of instant resocialization among Sudanese youth in psychological terms, we have to remind ourselves that most of the street boys had a good early childhood based on a stable and close relationship with their mothers and families. Many had a stable period in their early school years. When we followed the boys back to their homes, we found that their parents, neighbours and former teachers reported that many of the boys who left for the street were the brightest and the best in their class.

This helped us understand the street boys' confidence and initiative. Perhaps the boys who were insecure simply could not cope with the uncertainties and difficulties of street life.

Street children in Mozambique

Although researchers, relief workers and politicians related the problem of street children to the war situation in Mozambique, only a few of the children did. When asked why they left home, a majority told sad tales of family break-ups triggered by a father's departure, a remarriage, a new mother, or the beating and of favouring of other children in the family by the father, lack of food, and, no money for school-fees. Almost half of the children lost a close relative and many fled an area controlled and terrorized by the MNR and were not able to return to their home villages.

Almost all the street boys interviewed characterized street life as miserable.[12] They talked about themselves as lost persons, living in a terrible situation, mostly without a future. They missed their parents, siblings, relatives and home village. They were miserable during the cold season. They were chased away from the stores where they begged for money. They were beaten by police who also took their money. Smaller children were beaten and robbed by bigger boys. Many were increasingly organized by older patrons. They searched in the garbage for food. Many were weak and sickly. People shouted at them. They were accused of stealing. They worked hard whenever they could find a job only to be cheated out of their wages. "*The worst in street life is that your relatives pass by, but they pretend not knowing you.*" "*The worst to think about is that when you die, you die alone.*"

Asked about pull-factors and positive experiences, the children sometimes described some improvement over their old life such as more food, more money, or meeting nice people and friends. But moving to the capital had not necessarily offered more opportunities. On the contrary, many of them had to leave school when they fled and many would like to continue their schooling. Compared to the street children of Khartoum, the boys from Maputo appeared to be more miserable and thus perhaps more motivated for rehabilitation as their prospects on the streets were not as good.

The Mozambican boys actively responded to the hypothetical question: "*If God gave three wishes, which would be your first?*" Reply: "*A decent life as working members of the society,*" Most wanted to get off the streets.

Asked about the future, half of them chose to be a mechanic, while 20 percent said that they wanted to be a teacher. Very few chose powerful positions such as those associated with being a policeman or military officer.

The Maputo study identified, selected and recruited the most positive, reliable, verbal and mature boys who were then "promoted" to be collaborators in our study. We met with them every day and the researchers discussed the main findings that had emerged during the day's interviews. The group proved to be very useful, giving valuable input on topics like "street girls," organized crime, groups of street children in different areas of the town, and amounts earned (much higher than expected in the "good times") as well as on matters to do with war-affected children. They also recruited friends from the street with special backgrounds for further interviews.

Some important questions

Over the years we have met many street children, interviewing and interacting with them in their setting and at children's homes run by different organizations. We have followed them back home as returning "prodigal sons," but have also accompanied them as beggars, thieves, and daredevils, manoeuvering among the cars in fast traffic.

It was pointed out that the groups of street children in Khartoum, also called gangs, comprised members from all areas, all religions, speaking different languages, from diverse tribes. These groups functioned well together, in stark contrast to Sudanese society as a whole, which was torn apart by these same differences. The street children knew all about tribal animosities and, in response, called themselves the New Sudan. Is it possible that another and better reality could be born in the dust and sand from the actions of these street destitutes of the capital?

We made parallel observations among street children in Mozambique when we discussed the political situation with the street children there. Even the illiterate with almost no schooling were well informed about the history and contemporary situation of southern Africa. We felt quite confident that most of them were resilient, resourceful, with courage.

The boys who live in the streets of Maputo or Khartoum are more self-selected than driven. Maybe they have a stronger future orientation. Do they see the writing on the wall? One of the Sudanese boys fleeing the south with his parent's assistance described a good family relationship but added when asked about returning home: "*What do I get? Respect for religion and human rights? No. Education? No. Job, money, house, family? No. A bullet through my head? Yes!*"

Whatever their motivation, the children were willing to take risks to survive the turbulent urban life. We observed some extraordinary capacities in some of these boys, justifying an expression from one of the Sudanese street child workers who said: *"They are the explorers of their generation."* Others suggested: *"They get all their lazy muscles activated."* Street life was such a challenge that it required every bit of strength and initiative.

However, they are not all super children. They have lots of resources which are released and energized in the survival fight necessary on the streets. But they are also up against the wall. Even street children cannot live on dust and dirty words. The street is infected with violence, traffic, diseases and other life threatening dangers such as drugs and vice. If they fled the war area, did they land in a safer environment? We don't know the answers to all the questions, dilemmas and paradoxes but we firmly believe that the war-displaced child who is alone in the dangerous flow of city life, needs us. Our knowledge. Our admiration. Our respect. Most of all, our power to give them a safer voyage towards their adult future.

Most of the street children yearned for an education. In Khartoum over 1,000 had the chance to study in a technical school in the afternoon and evenings after the regular students were dismissed. Batch after batch completed the nine months training with flying colours and high marks for achievement and behaviour, doing better than those privileged enough to attend regular day classes. The instructors were reluctant to teach the street children at first. They feared that their tools and supplies would "go missing." But the street boys surprised them. They were more careful, more attentive and out-performed every expectation. When the course was completed, they were placed in private informal sector shops and businesses where they continued from strength to strength.

Are they the best and the brightest?

Having summarized our research in Mozambique and Sudan, based on the children's answers to questionnaires, to stories they made up to accompany drawings they were shown, sentences they completed, and their reactions to projective and indirect questions, we went on to recommend four guidelines for future research.

First is the push-factor. In the vast majority of street boy studies, the child in the street left home because of natural disasters such as famine, floods, drought and cyclone or due to the man-made scourge of terrorism and war. Despite this, only a fraction of boys from the affected areas, whether the terrorized provinces of Mozambique or the drought-stricken areas of Sudan's war-torn south, actually flee their villages and travel to the country's big cities.

Next come the pull-factors which attract the child and have to be seen in a double context. First, boredom in the home environment: for example, being left alone with siblings because the parents have to work 16 hours a day. Perhaps the children have the feeling they are a part of a society and culture that is at a standstill, offering them no prospect for the future. Second, children learn in the "dream factory" of the cinema about the freedom of westernized cities and the possibility of travel to other countries. Maybe this is symbolized by fantasies of easy money. More seriously, the lure of education, employment, and security attract children to the big cities. The pull appears stronger because of the very weakness of the pull-back forces which tie them to their family, village and roots.

A third viewpoint concerns family ties in times of hardship, such as drought and war and the concomitant lack of future perspective. Take the father who leaves his family in the Sudan to seek a job in another country. Very often, the father who leaves in search of work stays away so long that he finds a new wife and starts another family, forgetting about his previous obligations. Ties often break or at least loosen with long separations. Traditional customs such as bride prices and marriage arrangements which secure the family unit break down. Remarriage often occurs. A mother of four may have three more children by the new husband and if he leaves, she is left with a large family. Poverty, famine, drought, war and insecurity also put pressure on the family and in the last instance, result in the children being neglected.

Very often when we talked to street children who left home because of family quarrelling due to a bad relationship with a step-mother or father, rejection from the family seemed obvious. Some of the children ran away but still had some attachment for home and returned. Very often the child was disappointed, however, and, finding no welcome on returning to the village, thought he or she had been in the wrong to run away. But there have also been complete breakdowns of attachment where the child, very often the oldest, boldest or most independent boy, was thrown out or chased away by parents or guardians.

A fourth perspective is preparedness for the future. In interviews with children from the Sudan and Mozambique, we asked them about their future prospects. Although they had expectations for their personal futures with wives, children and jobs, they were much more concerned about the future of the nation and their region of the world because they understood that their future depended on the nation's capacity to handle the big problems. What was important for our purpose was to stress basic concepts that contributed to the children's decisions to leave home in our interviews. For the boys from war zones, the prospect of peace was of course one basic future consideration.

Maybe we are right in our observation that a majority of the street children we have met seem to be both brave and concerned about their future. These perspectives should be taken into consideration when intervention on behalf of these children is planned.

Only future research can provide further insight to the factors causing a child to live on the street. He or she is pushed and pulled and pulled back again in a tragic interaction within a family, usually under extreme pressure in a country in political turmoil and racked by war.

References

1. Dodge, C.P., Mohamed, A., and Kuch, P., "Profile of Displaced in Khartoum," *Disasters*, No. 4, Nov., 1987.
2. Mark, M.E., and Seligman, K., "City of Lost Boys," *LIFE* Magazine, June, 1988.
3. Gersony, R., "Summary of Mozambican Refugee Accounts of Principally Conflict-Related Experience in Mozambique," Report submitted to Bureau for Refugee Programs. State Department, Washington, 1988.
4. Loforte, A., "Criancas de rua em Mozambique." UNICEF Mozambique, 1991.
5. Dafeeah, E.N., "Vagrants in Khartoum. The Effects of Family, Schooling, and the Socioeconomic Status," University of Khartoum, 1986.
6. Ati, H.A.A., Children's Vagrancy in Khartoum: A phenomenon or a Crisis? University of Khartoum, Geography Department, draft, 1989.
7. Duffield, M.R., "The Blue Nile Lorry Trade: It's Nature and Relevance." In Manger, L.O. (ed.), *Trade and Traders in the Sudan.* Dept. Social Anthropology, University of Bergen, 1984.
8. Manger, L.O., "Traders and Farmers in Nuba Mountains: Jellba Family Firms in the Liri Area." In Manger, L.O. (ed), *Trade and Traders in the Sudan*, Dept. of Social Anthropology, University of Bergen, 1984.
9. Cederblad, M., Rahim, S.I.A., "A Longitudinal Study of Mental Health Problems in a Suburban Population in Sudan." *Acta Psychiatr. Scand.*, Suppl I., 79, 537-543, 1989.
10. Innstrand, A.G. Haaseth : "Street Children in Khartoum," Thesis University of Bergen, Norway, 1991.
11. Raundalen, M. : "Vagrant Children in Sudan," Report to UNICEF, Sudan, 1987.
12. Raundalen, M., Raundalen, T.S., Pereira, E., Vogt, N. : "Street Children in Maputo", Report to NORAD, Maputo, 1991.

CHAPTER 4

CHILD SOLDIERS OF UGANDA AND MOZAMBIQUE

Cole P. Dodge

Uganda: Background 1962-81

Uganda gained independence in 1962 under peaceful circumstances. Trouble started in 1966 when Milton Obote, in an attempt to consolidate his power, ordered his army to depose the King of Buganda, and made changes in the constitution. In 1971, Idi Amin, Commander of the Army, seized power in a military coup and ushered in a reign of terror and bloodshed which lasted until 1979 when the Tanzanian Army and a band of Ugandan exiles known as the Uganda National Liberation Army (UNLA) drove him from Uganda.[1]

Three successive weak governments followed until December 1980 when national elections, widely believed to have been manipulated, returned Obote to power. In early 1981, Yoweri Museveni formed the National Resistance Army (NRA), and led 26 men into the bush to mount an armed guerrilla war against Obote's UNLA. The NRA concentrated its struggle in the Luwero Triangle, a densely populated, fertile, coffee-growing region inhabited by 750,000 Buganda tribespeople just north of the capital, Kampala. During the next three years, the guerrilla war gained momentum and support. By 1983 the UNLA responded with a massive counter-insurgency campaign, which during a two-year period left an estimated 200,000 dead, drove 150,000 into UNLA-administered camps and displaced another 150,000 people.[2]

Child soldiers 1981-86

Against this background of civil war, thousands of homes were destroyed, communities dissolved and the fabric of social life in the triangle shredded. Families ran and hid in the bush to escape the atrocities of the predominantly northern UNLA soldiers. Parents and children were separated. Malnutrition and starvation escalated, proportionate to the UNLA's "seek- and-destroy" missions.

Meanwhile, the popularly supported NRA could not turn its back on the needs of the local people. According to the NRA, abandoned children were cared for in small numbers in the beginning, but by 1983 more and more school-aged children were absorbed into the NRA. As UNLA operations intensified and threatened remote NRA camps where these children were kept, a decision was taken to disperse the children and give them basic self-defence training to reduce risks. NRA officers "adopted" these children and looked after their food, clothing and shelter. After receiving basic self-defence training, the children soon accompanied the officers, carried weapons, ran errands, cleaned and cooked, and in this way became loyal contributors to individual officers and the NRA as a whole. They were highly motivated, reliable and dedicated, often instilled with a strong sense of revenge triggered by the UNLA atrocities against their families, friends or village which had driven them to the NRA in the first place.

In mid-1985, the NRA extended its sphere of control from the Luwero Triangle to Mubende, Fort Portal and Kasese. On July 27, the UNLA overthrew Obote in a bloodless coup and invited all "fighting groups" to join the newly formed Military Council. Smaller guerrilla groups plus the remnants of Amin's old army known as the Former Uganda National Army (FUNA) joined the Okello government (previously Obote's Chief of the Defence Forces). However, the NRA refused to join, seeing no prospect for any substantive change from the policies of repression of the former government of Obote.

As a consequence, the country was divided in civil war between the southwest one-third, controlled by the NRA, and the remainder, held by the UNLA, from August 1985 to January 1986. During this time the front line was a mere 20 miles from Kampala. Peace talks commenced in Nairobi and eventually a cease-fire was agreed in mid-December but never took hold.

The NRA achieved a greater degree of military success than it had anticipated, eventually encircling and isolating the UNLA barracks at Masaka and Mbarara and gaining control of the entire southwest. This long front line, spanning the area from Lake Albert to Lake Victoria and containing four major roads and the railroad as well as two large barracks still housing well equipped UNLA soldiers, meant that the NRA was spread thin and its resources heavily taxed. Every available soldier was pressed into action and by late 1985 this included the child soldiers. New recruits were rapidly trained and deployed and it seems that some of these were under 15 years. As fighting intensified in and around Kampala, more and more child soldiers turned up in the ranks of the NRA.

A major offensive was mounted on 17 January 1986, and Kampala fell on 26 January to the NRA. The child soldiers were suddenly highly visible as they moved about Kampala. They were the mascots of the NRA and became the subject of numerous newspaper articles and were frequently referred to as young liberators. To the credit of the NRA, they came equipped with a code of conduct and exercised a restraint previously unseen among the UNLA, the few young soldiers in whose ranks were especially feared. The NRA took prisoners of war and often assigned child soldiers to guard duty while the war front pushed slowly north. Child soldiers on road blocks and in the streets were seen to be reliable and

Panel 4

"I want the whole world to know!"
A twelve year old boy in Maputo finally escaped after 16 months of captivity. At ten years of age, he was snatched while playing; taught military skills and ordered to execute disobedient peers. He was then taken back to his village only to be forced to kill members of his own family when they fled their burning village. What message did this child veteran of atrocities have to tell us?

"Renamo destroys our schools and hospitals, they force the farmers to be refugees and destroy the nation and our future. Without the support of the apartheid regime of South Africa, Renamo would collapse immediately. We ask the whole world to help us in fighting the bandits. We want you to know that we suffer from destabilization all over the country and children are abducted and forced to kill, even their own families. I have seen it myself and I want the world to know."

His way of expressing himself was typical of how Mozambican children learn to tackle the psychological impact of destabilization through the Frelimo rehabilitation and political reorientation programme for child soldiers. Thus they have a frame for even the most horrible events taking place; a cognitive understanding of what is going on, based on political realities. This may prevent them from reacting irrationally.

Maputo, Mozambique, 1988.

trustworthy, seldom abusing the power of the gun which they all carried with pride and skill. Child soldiers were not allowed to smoke or drink in these early days following the capture of Kampala, just as it had been forbidden in the bush.

Uganda received considerable attention in the international media from the time of the July coup until mid-1986. Commentary on the child soldiers started soon after the country was divided in 1985. Feature articles glorifying the young soldiers appeared in the Nairobi press in late 1985 and early 1986.[3] BBC radio carried news of them as early as November 1985 (see chapter 6). Eventually, as Kampala fell and reporters had ready access to the child soldiers, feature-length articles appeared in major newspapers in the western world.[4] The BBC TV documentary, "Uganda: Children of Terror," appeared in late March 1986. Most of the western press reports carried some critical comment about the child soldiers while the Kenya press recognized the positive role the NRA had played in initially protecting the youngsters and gave credit to their role in the liberation of the country.

UNICEF opened a dialogue with NRA leadership on the issue of child soldiers on United Nations Day, 24 October 1985, when the first "corridor of peace" flight landed in Kasese with medical supplies. Initial concern was followed by a three-and-a-half hour meeting between Yoweri Museveni, the NRA leader, and, on behalf of UNICEF, myself in 1986, when the entire issue of child soldiers was discussed. No agreement on their future was reached, but with mounting international pressure, the new government announced in late February that all child soldiers were to be removed from front line battalions and that children would not be deployed in front line action. The NRA also claimed that they had never pressed them into actual fighting and the International Committee of the Red Cross (ICRC) medical teams who provided front line medical care confirmed that there were very few child soldiers in the front line medical units on the NRA side before Kampala fell. While this is widely believed to be generally true, many journalists interviewed child soldiers who claimed to have engaged in fighting.[4]

The dialogue for the future and potential rehabilitation of the child soldiers was pursued principally by UNICEF until July 1986. The government announced that many returned to their homes in Luwero in May and, further, that all children in Luwero would receive free education. By June, it was estimated by the Prime Minister's office that 3,000 remained in the army, including about 500 young girls. Since we had never seen more than a sprinkling of girl soldiers, these estimates may have been high.

Discussion

Two principal options were available to the NRA. The first was to keep the children in the army and provide for their education within the military itself. Proponents pointed out that the child soldiers had no known parents and that their home had been the NRA. Further, they argued that the children themselves did not want to leave. Finally, it was stated that these children represented a potential to improve the quality of the NRA as they matured with credentials of loyalty, service and motivation.

The second option was to take the child soldiers out of the army and enrol them in civilian schools. The schools would provide a minimum of primary schooling plus a skill. Those favouring this option pointed out that the officers would be encouraged to continue to care for their "adopted" children. Simultaneously, a tracing process could be initiated to help identify and reunite them with surviving parents and relatives. By going into civilian schools, the children could be given the option of later choosing between a career in the army or normal civilian life. It was recognized that these children of Uganda's terror would carry with them the scars of violence and war for the rest of their lives. They were mature beyond their years, but the respect shown them by adults had more to do with the weapon they carried than with their involvement in liberation.

The precedent set in Uganda for incorporating children into the NRA in the first instance can be understood, but many Ugandans worried about the future. Some pointed out that Idi Amin and Basilio Okello had themselves been youthful recruits. If the child soldiers became dissatisfied with slow progress and continued injustice, would they form a radical clique to overthrow the government at some future time? Others wondered if, given the severe economic and political problems facing the country, the NRA could really keep track of the former child soldiers.

Some of the children would continue to harbour revenge motives into their adulthood, presenting unpredictable problems. We know from writings on children in Northern Ireland and the Middle East that child soldiers are deeply imprinted psychologically. Their embedded desire for revenge is believed to contribute significantly to the perpetuation of mistrust, hatred and a never ending cycle of violence.[5] In a 1984-85 study of 650 school children in Kampala, Masaka and Jinja, we found little or no revenge motivation - a very positive indication that despite all the violence and insecurity, these children have not yet internalized tribal animosities.[6] The child soldiers, however, were a separate and highly motivated group who were encouraged to hate the enemy (the UNLA or Obote's soldiers), predominantly from the northern tribes of Acholis or Lango. Since almost all of the child soldiers were children of Bantu (or southern) tribes and were predominantly Buganda, there may be seeds for future revenge against the north. Yoweri Museveni and other NRA officers felt that much too much "fuss" has been made about child soldiers by western reporters, diplomats and UNICEF. They pointed out that the Geneva Convention, which prohibits children under the age of 15 to bear arms, was written by predominantly western countries and that it does not apply to Uganda's situation. They maintained that training in the use of a gun is just an extension of traditional values, and that even for the child soldiers who have experienced front line action, there is no basis for concern about psychological scars and after-affects.

While it is true that children in Africa do "grow up" sooner than their western contemporaries, in terms of responsibility for contributing to the family work and income, it is still not at all clear that these child soldiers have the maturity to handle the stresses of a war experience. Other Ugandans believe that the impact of war leaves long-term psychological problems for a significant number of combatants and that these problems would likely be magnified in the case of child soldiers. Regrettably, we were not able to survey or study the child soldiers in Uganda and we do not know of any other studies. Because Uganda decided to keep some children in the NRA while releasing the majority (but with no register) it is now extremely difficult to conduct follow-up research.

Child soldiers are increasingly deployed in guerrilla and conventional wars in developing countries, yet we have little or no information on how this affects them. The Uganda situation, however commendable from the NRA perspective having responded in humanitarian fashion to the destitute children, set a negative precedent in light of world opinion which condemns induction of children into military service. Since our research among ordinary school children found almost no revenge motivation, the questions are prompted: "Were the child soldiers a self-selected group who held revenge?" "Were they destitute and separated from their families or did they leave home with the intention of joining the NRA?" In retrospect, we worry that the number of child soldiers within the NRA may have been much more significant than was earlier thought. However, we note that the United Nation's Convention on the Rights of the Child specifically prohibits recruitment of children below the age of 15 into any army and that President Yoweri Museveni attended the World Summit for Children and ratified the Convention.

Mozambique : Background 1975-80

The recruitment of children into the MNR in Mozambique is done by capture, intimidation and brute physical and psychological force. Their experiences are horrifying by any standards, and even more revolting when the background to the civil war is understood. The Frelimo government came to power in 1975 following a long war of liberation from the Portuguese. Frelimo mobilized the vast majority of the population to their popular cause. However, neighbouring Rhodesia took offence at the newly independent Frelimo government because of its support for an independent Zimbabwe. Therefore, Ian Smith supported the creation of the MNR to destabilize Mozambique.[7]

With the overthrow of the white regime in Rhodesia in 1980 full support for the MNR was assumed by South Africa.

South Africa's destabilization policy in the front-line States, especially Mozambique, is well documented.[8] This policy included not only an economic blockade but, more devastating, arms and ammunition support to the MNR as well as the supply of sophisticated communication equipment. Training bases inside South Africa manned by military and secret police trained the commanders and fresh recruits from Mozambique. However, Mozambique fared well in the post-independence period and did not feel the MNR impact until after the formation of the Southern African Development Coordination Conference (SADCC) in 1980. It was not until the mid-1980s that the destabilization policies of South Africa peaked in Mozambique.

Child soldiers: 1980-90

A typical MNR recruitment scenario follows these lines: A village is attacked and looted; anyone who resists is killed on the spot. Men, women and children are forced to carry looted goods. Once they have reached an MNR base, often distant from their home and inhabited by a different tribe with a different language, the captives are forced to construct a crude hut and to work in the fields to produce food for the MNR. Women and girls are routinely raped and abused by MNR troops. A small number of boys and young men are separated from their families and trained as guerrillas. As the boys are subdued, they are taken on night raids into villages and if they perform well, given a gun and ammunition. The initial process is easily understood from a direct quote from a 15-year-old interviewed in 1989. *"They lock them 2 or 3 days in a house then they train. They give you a gun. After they give you a gun you are not allowed to run away. Because if you run away the bandits - the bandits they know where you live and they go and look for you. And they kidnap or they kill all your family. Some children are afraid of running away because they know if they run away their family will die"* (see chapter 5).

The strategy of the MNR was to intimidate young boys to the point where their socialization pattern was broken down and they actually accepted a gun willingly. Then they were forced to kill someone. The typical process entailed taking the boy back to his own village and forcing him to kill someone known to him. The killing took place in such a way that the community knew that he had killed, thus effectively closing the door to the child ever returning to his village. Over a period of time, it seems likely that the boys would sympathize with their captors. A well known psychological adaptation takes place wherein captives kept in close confinement develop a dependency relationship and eventually identify with the cause of their captors. This seems likely to have happened in the case of child soldiers in Mozambique.

As conditions have deteriorated, reports have circulated that extreme food shortages and repeated raids on villages have driven a limited number of people including youth into the MNR camps in search of food and protection.

While no reliable data are available, it is estimated that 8,000 to 10,000 children have been inducted into the MNR. Some 300 to 400 of these child soldiers have escaped or been captured and now live on the Frelimo side. Thirty-seven of these have been extensively interviewed by the press and their stories widely published.[9] The brutality these youngsters have experienced might be unbelievable were it not for the scars they bear, their missing fingers or slashed ears.

Note

Few, if any relief agency representatives have been able to meet the MNR leadership inside Mozambique let alone establish any kind of direct link with these child soldiers. While the Frelimo government is genuinely supportive and undertakes to rehabilitate, re-socialize and provide education for the ones who manage to escape, no longitudinal studies have taken place to evaluate the longer-term psychological effects on these young soldiers.

An earlier version was published in *Cultural Survival Quarterly*, Vol. 10, No. 4, 1986.

References

1. Avirgan, T., Honey, M., *War in Uganda: The Legacy of Idi Amin*, Tanzania Publishing House Ltd., Dar es Salam, 1982.
2. Johnston, A., "The Luwero Triangle: Emergency Operations in Luwero, Mubende and Mpigi Districts," in Dodge, C.P., Wiebe, P.D. (ed)., *Crisis in Uganda: The Breakdown of Health Services*, Pergamon Press, Oxford, 1985.
3. Amin, Mohamed, "Children of Terror," magazine section: *The Sunday Nation*, Nairobi, March 16, 1986.
4. Harden, B., "Political Violence Has Left Psychic Scars on Ugandan Children," *The Washington Post*, Washington, July 15, 1986.
5. Rosenblatt, R., *Children of War*, Anchor Press, Doubleday, New York, 1983.
6. Dodge, C.P., Raundalen, M.(eds), *War, Violence and Children in Uganda*, Norwegian University Press, Oslo, 1987.
7. Moroney, S. L., *Africa Handbooks To The Modern World* Facts on File, New York, 1989
8. UNICEF, Children on the Front Line, UNICEF New York, 1987.
9. Frelimo, *"The Children of War,"* Ministry of Information, Maputo, No. 7, 1987.

CHAPTER 5

INTERNALLY DISPLACED:
A SILENT MAJORITY UNDER STRESS

Rune Stuvland and Cole P. Dodge

Background

These people are a silent majority because they have no special rights under the provision of the United Nations compared to their compatriots who cross an international boundary and become refugees.[1] When they flee civil war, they flee insecurity in search of peace and a means to survive.

The internally displaced in Uganda lived in the centre of the country, near Kampala. They did not cross a border but nonetheless suffered harassment, torture and forced dislocation to camps at the hands of the government army, the UNLA. Up to 500,000 of the 750,000 inhabitants of the Luwero triangle were displaced and were provisioned in camps.[2] They lived with food scarcity, grossly inadequate shelter, no water supply, no schools and inadequate and sporadic health services provided by the ICRC, Uganda Red Cross and Save the Children Fund (SCF) U.K. An estimated 50,000 to 200,000 civilians were killed compared to 2,000 soldiers in the same period.[3,4] The IMR, was found to be 305 per 1,000 live births compared to 102 for the country as a whole.[5] These displaced lost their rights, their security, their homes, their livelihood, and their dignity.

In Sudan, the internally displaced have run the gauntlet of hostile armies, militia and antagonistic ethnic groups (see chapter 1). When and if they reached the relative safety of the capital, Khartoum, they faced continued deprivation, hardship and suspicion by the northern Arabic inhabitants. What little work they found was as day labourers at starvation wages. The government was at first unconcerned, then embarrassed and finally disgruntled by their presence. Land was not allocated to them in the "master plan" of the city. Services such as water, health centres or sanitation were not sanctioned by the municipal authorities to enable United Nations agencies and NGOs to help them with the most basic services such as hand pumps for their water supply. It was not until 1989, four years after hundreds of thousands had arrived from the war in the south, that permission was granted to UNICEF to drill simple bore holes for hand pumps in and near the settlements.

Civil war erupted in southern Sudan in the summer of 1983 after an eleven year interlude since the post-independence civil war which lasted for seventeen years. By late 1984, the equilibrium of the region was severely disrupted and large numbers of people were internally displaced or crossed the Ethiopian border to become refugees. Displaced people increased to half a million in 1986 and further mushroomed to over one and a half million by early 1989 (see chapter 1). The majority of the displaced lived near Khartoum, with the towns along the front line buffer zone between the north and the south providing a safe haven for relatively smaller numbers. The rural areas across the rain-fed agricultural zone in the southern fertile crescent of northern Sudan absorbed large numbers as agricultural labourers.

The civil war was exacerbated by the Arab-African and Muslim-Christian dichotomy. The Khartoum government policy up to 1989 favoured Arab ethnic groups along the north-south border who were traditional rivals with southern tribals and therefore historically opposed to the southern-dominated SPLA. These groups were encouraged to form militia for their own defence against the SPLA. The government provided them with arms and ammunition. Well armed tribal militia escalated traditional rivalries and caused the displacement of primarily Dinka but also Shilluk and Nuer tribesmen from areas where tribal boundaries meet. This caused many women and children to move north. Boys and young men escaped to Ethiopia in search of safety. Consequently, 1.5 million were displaced in the north, another 370,000 became refugees in Ethiopia and an additional half million or so took refuge in the major garrison towns of the south. By 1990, it is estimated that half of the southern population has been displaced, either in the south, to the north or into neighbouring Ethiopia.

Fleeing the war or the famine induced by war entailed walking for long distances. Commercial transport services had long since stopped. Encounters with SPLA, cattle raiders, militia or the army, put the displaced in immediate danger. When they met militia or the army, they could be accused of being SPLA supporters. Conversely, the SPLA did not like them "defecting" to the north. Reprisals were the order of the day. If porters were needed by any of these adversaries, civilians were pressed into carrying supplies or herding cattle miles from their intended route. Older children were routinely detained and were thus separated from their families, often permanently.

Once the displaced were inside northern Sudan, they needed money to buy passage on "suk" (trader) trucks to the main east-west trade route or railway line. What little possessions they had such as jewellery, was sold or, if they were destitute, as most were, they were pressed into mat weaving or other menial work for the traders at low exploitative wages. If they could not manage the money for the passage further north, they were stuck in hostile territory at a remote outpost in the border area which was isolated during the rainy season with no food or work. Under these circumstances, many mothers were forced to sell their children to obtain sufficient money to travel further north. While visiting such

Panel 5

"Three times the bandits attacked our village. The first time the bandits came, my aunt was with me and my brother and we ran away and we hid in a cave and then the bandits took my brother. They killed my brother with knives. And my aunt... they cut the arms, they cut the tongue.

"The second time they took my other brother and they went with him to the bush and killed him. I didn't see him die, but the people coming from the bush escaping from the bandits came and told me that my brother was killed. When they killed my older brother, we ran away to my mother's place and then the bandits came into my mother's place and they took my other brother and they went with him and after a few days they sent him back. They didn't kill him.

"The third time they came and burned all the places. When the bandits came they divided in three groups; one group to the school, one to the village, and one to circle the village. We hid in the bush, they didn't see us so we escaped. I will never forget what I saw. When I remember, I used to cry. Sometimes I sleep without eating because I think of what I saw.

"When I think about the people being killed, I see like it is happening now. I see the pictures exactly like I saw the day they were killed. Sometimes I become like deaf people because I am hearing sounds like guns shooting and my ears become closed. Now I am afraid because it seems like the bandits will come once again here and take me. I dream that the bandits will capture me and go with me and cut me into pieces. I don't feel happy when I speak about this that's why I don't tell anyone. In school I have problems and I am thinking about what happened to me because I just become lost and I loose everything the teacher is saying and I start writing the wrong things. My wish for the future is to finish the war and to keep the peace; no more war. I hope not to see things like that happen anymore. I hope to have enough food, enough clothes and a job."

Annabella, age 10.

centres in southern Darfur and Kordofan, the police offered a registration book with names of the children and thumb prints or signatures of their parents as "evidence" that no slavery existed; the sales were voluntary hence legal in the eyes of the local police who were themselves northerners.

The further north the displaced travelled, the less severe the ethnic animosity and the more secure they felt. Seasonal jobs were available in the rain-fed fields of local farmers although wages were minimal. Because of the limitation on the number of displaced who could fill these relatively small niches in the agricultural market, the majority boarded the first available train or truck for the long journey north to Khartoum. Most travelled by train free or with minimally priced tickets.

The train trip was full of hazards. Train schedules were infrequent and erratic so that southerners crowded on to whatever train passed. Carriages were filled to overflowing with paying passengers forcing the displaced to sit on the roof. Water and food were in extremely short supply and temperatures frequently soared to 40 degrees centigrade. The journey took up to a week. Periodically, the sick, weak, very young and old fell off of the roof. They were assumed dead although a few survived, some crippled for life. Exhausted by the trip, some arrived in Khartoum only to die on the platform of the train station.

Once the travellers reached the capital, they qualified for a month's ration of cereal given by Sudan Aid (a Catholic NGO supported by WFP). But the struggle was not over as life for the displaced was extremely difficult even after reaching the capital. They clustered together in ethnically distinct communities within the already crowded urban setting although some lived as family units at construction sites. Forty distinct displaced communities had been identified in Khartoum, Omdurman and Khartoum North by early 1989.

One such community was Hillat Shook, home to 20,000 Nuer and Dinka from the Upper Nile Region of southern Sudan. A proud cattle-keeping people accustomed to rural life, these families crowded together in makeshift cardboard hovels built on a garbage dump on the outskirts of Khartoum. Why? Because no other land was available to them. Children played amid broken glass and rusty tin cans. Many families were headed by the mother who had to compete with sisters from the south who had arrived earlier. The women had to carry water from a standpipe one mile away as municipal services were not available to the displaced camps. The most destitute, gleaned sorghum from the lorry park four miles away. "*Merissa*" (sorghum beer) was brewed to sell illegally with the risk of severe Islamic *sharia* punishment when caught. Men worked as the lowest paid unskilled labourers in an already congested market. The government of Sudan in November 1990, moved the southerners out of Khartoum to a site 30 kilometres south. Too far from the city to work, their conditions had worsened. Aid agencies opposed the forced move saying that what small hold the refugees had on a livelihood was now gone.[6]

Mozambique

The problem of the internally displaced in Mozambique gradually increased and has been the primary concern of the government for the last ten years. Following its liberation from the Portuguese in 1975, Mozambique experienced positive economic growth and development until the mid-1980's. In retaliation for Mozambique's support for those opposed to his regime, Rhodesia's Ian Smith helped establish the Mozambique National Resistance (or "the armed bandits," as the Mozambicans refer to them), which, with South African support, sought to destabilize Mozambique.

After Rhodesia became Zimbabwe in 1979, MNR lost much of its power, although it retained some support from those in Mozambique who had lost their power in Frelimo's restructuring of the society. The reorganization and collectivism of agriculture was both unpopular and ineffective as was the way the *"curandeiros"* or spiritual leaders were removed in Mozambique and the dethroning of the *"regulos"* or local chiefs and tax collectors. But the potential for internal support decreased rapidly because of the extensive use of terror and atrocities which characterized MNR activities.

With support from South Africa, MNR intensified its efforts to destabilize the country from the early 1980's on. The security situation steadily deteriorated as the bandits used extreme terror and violence to deter the population, such as kidnapping and killing, and the MNR attacked railroads, roads, schools, health stations and commerce.

Large portions of the countryside were not cultivated because farmers were at risk of attacks and looting and transportation of food was risky. Communities were left with food shortages and inaccessible for relief distribution other than by air which was expensive at best and impossible at worst. Although South Africa signed an agreement in 1984 stating the cessation of its support to the MNR, the bandits were still active in all of Mozambique's provinces, causing insecurity and forcing people to become refugees in their own country. The displaced lived in camps as *"deslocados"* (displaced) and they still risked attacks from Renamo. The population of 15.6 million lost 3 million who were driven away from their homes as a result of food shortages and terror. Of these, 1.9 million were recently internally displaced, the remaining 1.1 million lived as refugees in neighbouring countries, more than one million in Malawi.[7] (It should be noted that the total number of internally displaced is much higher when the past 15 years are considered. Those who were displaced earlier are thought to have been integrated and therefore no longer qualified for food aid.)

The only study of war-traumatized displaced children, that we know of, is reported below although concurrent research was conducted by Boothby and other studies have been done.[8,9,10] The study provided part of the data for a thesis in child psychology at the University of Bergen, Norway, conducted by R. Stuvland.

The sample

In Maputo, we interviewed 14 displaced children between the ages of nine and 15 years. They were recruited for interviews through local officials in the two *"bairros"* of Motala and Patricie Lumumba with no particular attention to their socio-economic status. A class of 31 pupils was accessible in Patricie Lumumba. Our only criterion was that they must be internally displaced by the civil war.

We administered a one-page questionnaire to the students in order to screen them for war displacement. The purpose of this was to select those who were displaced and assess to what extent they had experienced the war. The results from the questionnaire revealed that half of the children were internally displaced. Twenty-five children in the class had been in a situation where they had to run and hide because of the war, and 19 had witnessed somebody being killed. Twenty-four children reported that someone in their family had been killed in the war, and one or other parent of six of the children had been killed. Further, 16 children were separated from their father and 10 from their mother because of the civil war. Seventeen children reported somebody close to them had been kidnapped or was missing, five children had been kidnapped by the MNR. In other words, not only the displaced but many others in the class had experienced the trauma of war.

Originally we planned to interview all 16 displaced pupils who had experienced severe war trauma, but because of transportation problems, this was impossible. Thus six displaced children were recruited for interviews -- those who lived near the school, enabling them to be interviewed after school hours. In Motala, a group of 15 children were accessible to the interviewer. Eight were over the age of nine, therefore recruited for interviews.

The 14 children interviewed in Maputo came from the three southern provinces of Maputo, Gaza and Inhambane. They had to travel distances ranging from 50 to 500 kilometres to reach Maputo. Some had left just a couple of months before while others were absent from their village for three years. Some stayed in Maputo the whole period, others had moved back to their village when the security situation allowed, only to be driven away again.

Eight of the children lived with their mother and father, three with only their mother and three with relatives. Nine of the children went to school in Maputo. This left five who either worked or who had left their identity papers, when escaping from the bandits, creating problems when applying for a place in school in Maputo.

The interview

A semi-structured interview guide was followed based upon the methodology developed for dealing with traumatized children. First, the children were given a free-drawing test. Second, they were asked to tell a story about their drawings and later, we focused their attention on their experiences of war and violence. Finally, we recapitulated what they had told and focused their attention on the situation of today and their hopes for the future.

The children were interviewed two to five times. Interview sessions lasted from 40 minutes to two hours. The number of times the children were interviewed depended upon the severity and type of trauma revealed during the first session, and the availability of the children for a repeat interview. For those children who had experienced repeated and severe traumas, we exerted considerable effort to obtain a comprehensive story. A coherent representation of the child's life from the first attack up to the time of the interview took time and entailed much more interaction than the normal standard questionnaire. Several of the children clearly had emotional difficulty relating their history to us.

The children were interviewed by Stuveland with the help of an experienced Mozambican social worker who acted as interpreter. Most of the children preferred to be interviewed in Shengana, the local language. Those who were interviewed in Portuguese often switched to Shengana when talking about difficult emotional matters.

The children's experiences

All of the children had experienced attacks by bandits on their home villages. Some villages had been attacked once, others several times. The usual experience was that the bandits attacked the village, looted and kidnapped people while most of the inhabitants escaped and ran away. Invariably some people were killed. *"The first time they came, they attacked the village, they stayed for a day and then they ran away. The family ran away from Moamba here to Maputo and when we heard the situation was calm we went back, and the bandits, after a few months, they came back, they attacked the village again..... They killed people, they took people with them, and burned houses and everything."*

Among the 14 children interviewed, nine had seen someone being killed, five had witnessed close relatives being killed. Several told us that they had hidden during the attack and from their hiding place had seen friends and relatives being caught and killed. Six of the children experienced attacks during which they witnessed mass killings. One child saw the massacre of as many as 58 people.

The brutality practised by the MNR towards the villagers in several instances, was extreme. Two children told of episodes where the bandits put people into a hut, tied steel-wire around the hut door and set it on fire. Others told of mutilation such as noses, ears or lips being cut off, and of people being killed by having their arms, legs and finally their heads cut off. Some had watched these atrocities happen to close relatives. *"I told them to run away with us, but they refused to run away with us, they stayed and hid in a hole. It was two aunts and one of my uncles with his son and one brother of mine. They stabbed my brother here in the front, they killed my brother with a knife. And one of my brother's cousins, they opened the stomach and they put the intestines on the chest of my aunt. They cut the arms, the legs and they cut the head off."*

All the children related that in connection with the attacks, the bandits kidnapped people to intimidate others so that they would act as porters to carry off looted goods. Four of the children were kidnapped. One of them had been sent home after reaching an MNR base camp. The three others escaped after a few days. The march took several days, and during these marches the children often witnessed severe harassment. If a person could no longer manage to carry the stolen goods, or if they refused, they were sometimes killed. Some of the children told how those kidnapped were divided into different groups. Some were sent home, others were recruited to "entertain" the bandits by providing sexual services, especially girls and young women, and boys and youth were coerced to be child soldiers or do other kinds of work. Different control mechanisms forced the captives to stay in the camp. *"I was kidnapped. And when you are kidnapped by the bandits, when the bandits take you, they make a small meeting and they kill some people while you are watching. You can't cry, you can't complain, because if you cry, if you complain they kill you. And they do it in front of all of us. So they started asking us many questions, many things, and in my group they chose five girls and the rest of us, they said 'You can go'."*

All the children had experienced separation from or loss of close relatives either or indirectly because of the war. Eight children had been separated from one or both parents for months or years. Nine had experienced close family members or relatives being kidnapped by the bandits, some of these were still missing. One child did not know whether his father or brother were dead, still held or living in another part of the country.

Another problem reported by the children was that of keeping in contact with family and friends living in their home area in the countryside. The roads and railway lines were vulnerable and often attacked, thereby restricting people's mobility. Besides the trouble created by separation, several children reported strong experiences of anxiety because their loved ones lived in districts frequently under attack.

The children's reactions

The nature of the study was an exploratory investigation of the children's experiences and reactions to war. We did not, therefore, use normative instruments. Reporting on the children's reactions and drawing conclusions from them for a larger group of children is not recommended. Nonetheless our findings are important in developing an understanding of the problems faced by the internally displaced in Mozambique.

The most common reactions reported by the children were their feelings of anxiety connected to what they had seen. They were fearful of new attacks and for relatives still living in less secure parts of the country. Their fears were especially obvious when we analyzed the children's dreams. Five of the children reported symptoms of re-experiencing traumatic events. Although not investigated through systematic questionnaires and scales, the interviews revealed that these children showed common symptoms of a post-traumatic stress disorder, as validated in standard psychiatric diagnostic systems (DSM-IIIR). The re-experiencing phenomena which children told us about were both visual and auditory in nature; either directly related to what they had seen during the attacks, or thinly disguised in the dream's manifestation of the traumatic events. Of particular interest was that children as young as nine reported re-experiencing traumatic events.

Reports of bandits passing through the village where the original attacks took place provoked flashbacks in all the children's reports. The flashbacks usually contained one or two distinct episodes in which the children were very vulnerable and were associated with the violence they had witnessed. A typical example of this was a situation in which the child hid from the bandits, and watched someone being mutilated and killed. In some cases, the person killed was a close relative. These episodes were "replayed" in the children's flashbacks again and again, leaving the child extremely vulnerable as he or she could not escape the feelings of terror so frequently stimulated as the war went on and on. *"I see like it is happening now. I see the images exactly like I saw the day the people were killed. And sometimes I become like deaf because hearing sounds like guns shooting and my ears become closed. When I remember that, I used to cry sometimes, and I can't eat, sometimes I sleep without eating because I think of what I saw, I have never seen things like that."*

Flashbacks also interrupted the children's concentration, and were particularly problematic for those attending school. Some of them told us that they missed what the teacher said and wrote wrong "things" as these flashbacks occurred. *"I have problems when I'm concentrating about what happened to me, because I just become lost, I loose everything that the teacher was saying and then I start writing the wrong things."*

Sleep problems reported by several of the children, were most extensive in the first days or weeks after the traumatic episodes. But a few reported that not being able to sleep was a big problem even at the time of the interview, which, for some, was two years later.

Talking about their experiences

The children who experienced the worst traumas were the ones reporting the most severe psychological symptoms. All the children reported that they seldom, if ever, talked about their experiences with anyone else. And although the interviews activated memories and feelings connected to their personal exposure to violence, they reported feeling "good" after talking to us. This was the case even for those who had experienced the most severe, repeated traumas and were initially reluctant to talk about their experiences.

Teachers knew very little about their pupil's history. They seldom talked about the children's private experiences of war and violence. It seems plausible that the teachers, besides lack of knowledge about the importance of speaking to children about traumatic experiences and how to do this, actually avoided these subjects because they were upsetting for the teacher, and for the child. We confirmed that although it was upsetting for the children to initially talk about their experiences, it made them feel better in the long run provided they were allowed to talk through their traumas in a safe and supportive environment.

When a child talks about his/her experiences, one should not avoid even the most brutal or inhuman aspects of the story, but support them in expressing all the details, however gruesome. In this way, he/she learns that a responsible adult is able to cope with even the worst stories of violence; thereby assuring the child he/she can cope as well. If the adult avoids the child's experiences, the child learns that what has been experienced or even done (thinking here of the child soldiers) is too horrifying to cope with, thereby confirming the child's anxiety. The children need to be given several opportunities to talk about what they have experienced.

Discussion

Silent majority fits our understanding of the internally displaced in the three countries under consideration. Silent first because they have no rights similar to those classified as refugees by the UNHCR. Second, because the relief and emergency assistance provided by UN, bilateral donors and NGOs often cannot reach the displaced either because of insecurity or because government authorities do not allow access (or sanction permission) for such assistance. However, in Mozambique there is a specially designated government department to cater to their needs and the United Nations Development Programme (UNDP) has recently taken some initiative and leadership in this area. Third, because they are dislocated within their own country, it is sometimes difficult to distinguish them from the local population. Without political influence and far away from home, the displaced are more an inconvenience than a priority for development planners as their future is unpredictable and thus thought not to be worth an investment. These planners overlook the human potential and development of skills which will facilitate integration and resettlement when the conflict ends.

Special problems

- School children in Maputo had to flee their homes so quickly that they left their identity papers behind or their homes were burned by MNR guerrillas and the children were subsequently denied school admission.
- Rosa, the fully qualified school teacher who lived in a hovel at Hillat Shook in Khartoum with hundreds of school-aged children living all around her, could not teach because her provincial government was bankrupt. The central government had stopped all salary transfers to civil servants because the schools were closed in the south. She brewed illicit beer and sold it illegally to keep food to eat on her mat on the floor of the garbage heap for Khartoum.
- Edward whom we met with a head load of bananas trekking from the wholesale market in Kampala, worked as a day labourer after his homestead was overrun by UNLA soldiers. His bicycle and other household items were looted, forcing him to work for the first time in his life as a labourer. He had once owned his own gardens and coffee trees.
- Or the Sudanese refugees in western Ethiopia who were visited by the first international delegation who "...*believed that there were some fifty cases of severe trauma among children between ten and twelve years old - a grim testimonial of the ordeal of the march from southern Sudan.*" When they visited the clinic "...*it was more than most people could deal with without severe emotional reaction...several members broke down and cried.*" as reported by Ray Bonner in his article "Famine," which appeared in *The New Yorker* (13 March, 1989).

These are the war displaced, the individuals who live under continuous stress and insecurity: the stress of fleeing; anxiety about their homes; separation from family, kin and neighbours; broken families; unemployment; thrown into camp life without schools ..... resulting in grave uncertainty about the present let alone the future. Yes, these children are the silent majority in the conflict situations of Africa.

References

1. Dodge, C.P., Mohamed, A., Kuch, P., "Profile of the Displaced in Khartoum," *Disasters*, No. 4, Nov. 1987.
2. Kajubi, Senteza, "Background to War and Violence in Uganda," in Dodge, C.P., and Raundalen, M., ed. *War, Violence and Children in Uganda*, Norwegian University Press, 1987.
3. See chapter 1, table 2.
4. U.S. Committee on Refugees. Newsletter, 1985.

5. Dodge, C.P., Henderson, P.L., "Recent Health Survey:" Towards a Morbidity and Mortality Baseline," In Dodge, C.P., Wiebe, P.D.,(ed) *Crisis in Uganda: The Breakdown of Health Services*, Pergamon Press, 1985.
6. Perlez, J., "Sudanese Troops Burn Refugee Camp," *The New York Times*, 1 November, 1990.
7. Green, R., Asrat, D., Mauras, M., and Morgan, R., *Children on the Front Line*, UNICEF, New York, 1987.
8. Pynoos, R.S. and Eth, S. "Witness to Violence: The Child Interview". *Journal of the American Academy of Child Psychiatry*, 25, 306-319. 1986.
9. US Committee for Refugees, Support for Mozambican rebels should cease, *USCR*, Washington D.C., 1990.
10. Cliff, J. L., and Noormahomed, R., "The Impact of South African Destabilization on Maternal Child Health in Mozambique." *Journal of Tropical Paediatrics*. 34(6), 329-330, 1988.

CHAPTER 6

CORRIDORS OF PEACE ACROSS THE LINES OF CIVIL WAR IN UGANDA AND THE SUDAN

Cole P. Dodge

Background: Uganda 1985-86

"I wished I had poison. Military men came and camped in our area. They began harassing innocent civilians. They began killing people. They used to kill using knives. Others were ordered to pierce themselves with thorns" (13 year old girl from the Luwero Triangle in Uganda, 1985).

Children born in Uganda since 1970 accounted for nearly half the country's population in 1985. All of them have grown up in a country that has known dreadful violence. Many have directly experienced personal loss and assault. Ugandans naturally looked to their children as their "hope for a better world," their resource for a more constructive future.

Negotiating to reach children caught in war is difficult. Primarily because civilians are one of the major targets. Secondly, the news media is a significant part of tactical planning and each side may perceive that civilian suffering helps their cause. Relief assistance is not necessarily a priority for either side and where it is attempted without their agreement vehicles delivering relief supplies may be specifically targeted. Convincing the belligerent sides in a civil war is a difficult, complex and risky business. There are few successes to draw on as precedent and all make the case for a role for international groups such as the International Committee of the Red Cross and Médecins Sans Frontières (MSF or Doctors Without Borders).

UNICEF helped mobilize the Uganda media to publicize the impact of war on children in 1985 when the corridor of peace was established to take relief to those outside government-controlled areas. Questions such as "Are you satisfied with Uganda today?" and "Do you want to leave your children with a better tomorrow?" were addressed to the public. A six-part documentary produced by Uganda television entitled "Give Children a Chance," was supported financially by UNDP and endeavoured to convince the public that more children were at risk of death from vaccine-preventable diseases than from bullets. Ugandan radio and TV hammered away at the theme that each child had a "right" to immunization regardless of what side of the war they were caught on.

Security had completely broken down in Uganda by mid-1985. President Obote's disgruntled army seized power, dividing the country between government and rebel NRA areas.[1] UNICEF persisted in its efforts to reach children across the lines of civil war and especially to reach those in the rebel-held areas where country-wide programmes to improve health care had ground to a halt. UNICEF helped organize a two-day seminar which brought Uganda's religious leaders together to review "Give Children a Chance" in an attempt to marshall public support for reaching children caught behind the front lines. Attention was focused on the impact of war on the society and on the children as innocent victims. The mobilization of religious leaders and their constituencies helped to develop a general public consensus. This and the television coverage greatly contributed to the success of the corridor of peace.

Corridor of peace

What was the corridor? Where did it begin and end? Who established it? How? Why? We looked at and considered many options. Road transport was the most logical. For the first time in the three-year civil war, we had a clearly demarcated front line and we knew the relative positions of both sides. Since cost is always a consideration, we thought a road convoy of 30 clearly marked Red Cross and UN trucks would be our first option. Failing that, the only way to access the rebel-held one-third of the country was by air. A number of possible air fields were identified as landing sites.

To support the concept of the corridor and to ensure access to the children, we cited the legitimacy of the United Nations and brought the ICRC into the planning. Examples of UNICEF's own work in wars ranging from China to Central America and the Middle East along with the long and distinguished history of the ICRC were cited. But even then, when confronting the various ministries and departments of a sovereign government, we needed a specific legal frame of reference. UNICEF's Basic Agreement with the Government provided for "access to all needy children" as well as the freedom to "assess the conditions and publish findings on children throughout the country." Only the Ministry of Foreign Affairs was even vaguely aware of this document and the Geneva Conventions which governed the ICRC mandate, yet without these we could not have persuaded the civil and military authorities. As far as I am aware, this was to be the first time that relief supplies would cross the lines of civil war from within a country with the consent of both sides. Usually, cross-border relief operations are mounted from a sympathetic neighbouring country: occasionally the govern-

Panel 6

"Myself in the year 2000"

"I am a girl of 16 years. I go to school. And when I am finished, I will go to the University of Juba. I will be typing. But not here in Juba. Outside. I will go and work there because I want to help my parents and my brothers and sisters. But when this war does not stop, I will take them outside, because when they sit here, I will not be happy. But now I pray for God to help us here. When we enter in that time and the war is stopped, I will build a large house for my mother and father. Because for that time, they will have no power to work. But for that time, I will take care of them. Because now my father tell me, 'Rebecca, my girl, you try to read very hard because in your future no one will help you, but you will help yourself and us also.

"I want to be a (Catholic) sister. I do not want to marry, because when I marry, who will help my mother and father?

"But now I want to go to school and when the teacher says I shall pay my 50 (Sudanese) pounds, I tell my father and he gives me because he knows when I will finish school I will help them. And this is the end of the story about myself."

Essay from Juba, the Sudan, 1989.

ment in power may turn a blind eye or agree under extreme international pressure to cross-border relief operations, but, more often, they protest these clandestine arrangements. The precedent in Uganda was hard won.

Hard won in the sense that the NRA told me: *"Yes, we want your help but bring relief from across the border from Tanzania, Rwanda or Zaire."* The government in Kampala said: *"We will not give medical aid to our enemies."* Why risk being expelled at worst or placed in official disfavour at best if authorities in Kampala found out, as they surely would if a cross-border relief operation was launched? Setting the precedent for a corridor of peace within the country was a much better alternative. Several lines of negotiations were used: for example, reassuring the government that UNICEF had dealt only with the NRA in the context of local authority and not as an alternate government. One convincing point with both sides was the principle of giving assistance only to civilians, with the emphasis on children, something that UNICEF and the ICRC were adamant in enforcing and ensured that the government and NRA alike were aware of. For example UNICEF followed up vigorously whenever a vehicle or supplies were hijacked, stolen or misused. It helped too, that UNICEF was trusted because of its vigorously implemented and successful water drilling and health centre rehabilitation programmes which continued despite insecurity, inflation and such marketplace scarcities as fuel supplies and shortages of materials.

While government negotiations were always conducted face to face, often with the army Chief of Staff but also with civil servants in the Ministry of Foreign Affairs, the corridor of peace negotiations with the NRA were made by telephone between their headquarters in Nairobi and my office in Kampala. Although I was sorely tempted to meet them on one of my frequent trips to Nairobi, the only personal contact took place on Ugandan soil after the corridor of peace was established in NRA territory. This, I believe, was significant in overcoming pressure from the NRA to mount cross-border relief assistance. Had they managed to get me to meet them in Kenya, my principle of dealing with the Ugandan problems within Uganda might have been compromised. Also, my reluctance to comply with their demand for a cross-border operation generated sympathy when I explained we risked expulsion from Uganda if the organization provided relief without government approval on a cross-border basis. The government would not agree because they would not have any idea of what was being provided. James Grant, Executive Director of UNICEF, fully supported these efforts to establish a precedent-setting corridor of peace for children.

One of the first and most difficult hurdles was to obtain the necessary travel approval to visit the rebel area from the Ministry of Foreign Affairs, since all UN international staff needed permission prior to travel outside of the immediate vicinity of the capital, Kampala. Eventually, after citing the Basic Agreement to the effect that the government allow access to all children in "need" and suggesting that this might have to be renegotiated, which would entail a temporary curtailment of all assistance, the Ministry of Foreign Affairs agreed to a compromise--a special disclaimer would be written into the travel authorization to protect the Chief of Protocol who had to sign it. The approval was couched in the following wording to absolve him and his government if anything went wrong: *"We advise you not to go to the conflict zone because we cannot guarantee your safety but at the same time, we do not deny UNICEF staff access to all parts of Uganda to serve the interest of needy children."* Olara Otunu was the Ugandan Permanent Representative to the UN prior to returning to Kampala as the Minister for Foreign Affairs. He knew UNICEF well and personally gave his support which carried considerable influence with the government. The historic credibility of the International Committee of the Red Cross was also used as a strong point in the negotiations through a local agreement wherein front line emergency medical services were provided based on the ICRC's mandate to provide aid and protection on the battle field. Although UNICEF took the lead in negotiations and provided behind the lines assistance, flight costs were shared equally between the ICRC and UNICEF. In addition, the psychological advantage of negotiating as two agencies allowed for a "fall back" position whenever more time was needed by saying that the ICRC had to be consulted.

First flight

The role of the news media should not be underestimated either. For example, in an interview with the BBC, I announced that both the NRA and the UNLA, had allowed five flights across the lines of civil war to immunize children. The BBC World Service carried the story as a lead item which was heard around the world but most importantly on both sides of the war in Uganda. The government's reaction was to accept credit and state on TV and radio: *"We have been encouraging UNICEF to take supplies to the rebel area, after all, we all have children and we are all Ugandans."* This was the very argument we had employed to convince them in the first place and religious leaders were fully supportive of this approach. Also, media professionals were convinced that a corridor of peace should be opened and sustained. They went on to explain that the vaccines carried by UNICEF to rebel territory could only be used to protect small children and could not be diverted to combatants.

The first flight across the lines of war took place on United Nations Day, 24 October 1985. Negotiations with the NRA in Nairobi had been in process for about two weeks and two earlier dates had been fixed for initial flights. Recognizing that this was a risky affair, these dates were arranged so that the reliability of the Nairobi contact could be verified. It would have been a disaster to fly into a situation where our aircraft and UNICEF/ICRC officials might be at risk. Even though the NRA high command had agreed, it was imperative that we know that the ground troops were expecting us. The regional office of UNICEF in Nairobi was supportive throughout, although the officer-in-charge, Ken Williams, requested an appraisal of the relative risks. Being a statistician, he wanted assurance that all angles had been adequately thought through. To ensure arrangements for a safe landing in NRA territory, I radioed the Ugandan staff of the UNHCR in Fort Portal on the NRA side, requesting them to drive to the landing strip and confirm that the local NRA commander and the officer in charge of the unit securing the air strip were expecting a flight. Having confirmation, I scrambled to put together a joint UNICEF/ICRC relief team of international staff, including Araya Assefa as the UNICEF staff member, who set off in a convoy from Kampala through north east Uganda and crossed the border into Zaire. Skirting the western side of the Rwenzori mountains, they eventually re-entered Uganda and arrived in time to meet the flights. These staff kept our communication lines open and were essential to the integrity of the corridor of peace as they assured proper use of the supplies. The ICRC had their own team based in Kasese which was able to assist with flights as well as provide front line emergency medeical care while the UNICEF team was able to help distribute vaccines and other medical supplies to government and mission health facilities.

Since neither UNICEF nor the ICRC had an aircraft based in the country, a Kenyan-registered plane operating out of Entebbe which was leased to UNHCR for refugee work in north east Zaire, was chartered for the first flight. A flight plan was filed with the Entebbe Civil Aviation Authority for a flight to Arua in West Nile (controlled by the government) with an intermediate stop in Kasese (under rebel NRA control). When the airport control tower received the flight plan, the controller called the pilot and myself to the briefing room. He demanded a full explanation concerning the flight. I showed him the military clearance for Arua and the Ministry of Foreign Affairs travel permit for Kasese. He was clearly unhappy about this clearance but after twenty minutes of arguing, we parted. No mention was made of Kasese when the controller cleared us for take off. There was no clear permission from the military so we took care to assure a fast intermediate stop in Kasese with just enough time for me to disembark and off-load the 500 kilograms of emergency medical supplies. The pilot then flew on to Arua and picked me up on the return flight while the air traffic controller in Entebbe chose not to make an issue over the request for an intermediate stop in Kasese...granted, the pilot did not report the intermediate stop on the radio to the air traffic controller (technically a breach of flight procedure) but nonetheless, we established the precedent and established credibility with the NRA.

With one flight conducted safely, I went to the tower when we touched down in Entebbe and thanked the controller for his help. He was full of curiosity about the NRA and pleased that I shared my first impressions with him. In this way, he shared the success of the venture and felt he was a part of achieving something positive for children as I took the time to explain to him what we were doing. Over the next few days, we filed a flight plan for Kasese using a different plane which belonged to a local Ugandan-registered company. This company had a standing or blanket clearance to fly anywhere in the country issued by the former Minister of Defence, Tito Okello, the then Head of State. The controller was much more comfortable with this clearance. The local company had confidence in our arrangements with the NRA after the success of the first flight and wanted to establish their own credibility with the NRA as a hedge against the future. Again, the Ministry of Foreign Affairs travel permission for UNICEF staff and the military clearance for the aircraft did the trick with the airport authorities.

Corridor at risk

But military authorities soon found out about the mercy flights. Once aroused, they were suspicious, especially once a military helicopter defected and landed in NRA-held territory with ammunition and crew. The loss of the chopper on the day of our fifth flight was clearly too much for the military. Basilio Okello, Minister of Defence, notorious for his hot and erratic temper, summoned me. En route to his office, I called on the Chief of Protocol at Foreign Affairs and the UNDP Resident Representative and told them of the summons as some insurance against being imprisoned or deported. Armed with nothing more than a healthy fear and official documents such as the signed Basic Agreement dating back over two decades and the Ministry of Foreign Affairs travel authorization, I reported to army headquarters. Escorts took me straight into the minister's office and there, despite the tirade aimed at me, I managed to make it amply clear that the flights were valid and legal both within the context of UNICEF's mandate for reaching children and the law of the land (at least from my perspective). Finally, when all other lines of reasoning failed, I marshalled every ounce of conviction I had and despite a cold fear which crept down my spine, suggested that it might be necessary to stop all UNICEF assistance and evacuate international staff if the flights were not allowed to go forward based upon a deteriorating security situation. Basilio Okello knew that his government was in disarray and suffering from more international disfavour than the Obote government which had just been overthrown. Bilateral aid had already been reduced and some UN system technical assistance personnel had been evacuated, so my suggestion was an effective ploy. UNICEF was one of the few agencies still working in Kitgum District, the home area of the Okellos. No clear new permission was granted, and the meeting ended in a stalemate. It was clear to me that the government was upset, that my own security might be in question and that the continuation of the corridor of peace might be in danger. The following day, another flight to Kasese was attempted but met with a solid wall of non-cooperation. The central medical store staff refused to release UNICEF-supplied emergency drug kits and the airport staff (especially the military security) would not even entertain the idea of another flight to Kasese. By early afternoon, I decided to take my family to Nairobi for the weekend fearing that we might be in danger if we stayed in Kampala. (We lived just up the hill from the notorious Mbuya army barracks.) Up to this time, no other UNICEF staff based in Kampala was directly involved in negotiating with the authorities; however, Sally Fegan-Wyles, Chief of the UNICEF Health Programme, was fully appraised and kept the Ministry of Health in the picture. Just prior to taking off from Entebbe, the nervous airport security forces opened fire on one of their own helicopters fearing another defection. Since our Kasese flights were widely talked about among the security forces guarding Entebbe and because we were flying in the same aircraft even though the flight was cleared for Nairobi, we flew just above the long runway as a precaution and skimmed over Lake Victoria until we were out of range of the anti-aircraft guns. Mike Woldridge, BBC East Africa correspondent, arrived at our Nairobi hotel minutes after we checked in and asked for an interview. My movements were fairly widely known as the Kasese flights created considerable media interest. Relieved to be safely in Nairobi and pleased with the first five flights and acutely aware that all was lost unless a media wave was generated, I talked openly to the BBC. The interview concentrated on questions concerning when and where I had been in NRA territory and what I had seen. The fact that the flights had been undertaken with the consent of both the government and the NRA, and had, as such, established a precedent for a corridor of peace across the lines of civil war inside the country as distinct from cross-border operations was emphasized. The account of my impression of landing in Kasese and the large number of NRA child soldiers there was also highlighted.

The interview was carried by the BBC as a main news item and featured on the African news. Heard by government officials in Kampala and NRA officers in the "bush" and their high command in Nairobi, it prompted the Okello government to publicly take credit for the flights while simultaneously, using my comments about the child soldiers to criticize the NRA on their own radio and in Kampala. They understood my interview to be pro-government, which helped pave the way for more corridor of peace flights.

Corridor secured

Forty-six flights originated from Entebbe and flew with supplies released from government central medical stores for delivery to government and mission health facilities in rebel-held areas. Pre-addressed medical kits were delivered to each of the 109 health centres throughout the country. Those health facilities on the other side received kits through the corridor of peace. This was a routine part of the UNICEF-supported health centre rehabilitation programme. Vaccines and oral rehydration salts (ORS) were also supplied and carried by chartered aircraft from Entebbe to Kasese through the corridor.[2]

It was fortuitous that I reacted to the young boy soldiers guarding the landing strip by pointing out to the NRA officials on the spot that the use of child soldiers was contrary to the Geneva Convention. Consequently, the NRA could not accuse me of speaking "out of turn" to the BBC. Later, when I met NRA leader Yoweri Museveni in Masaka, before he captured Kampala and became Head of

State, we discussed the child soldier issue at length. This gave me the confidence to continue to speak openly about child soldiers to both the government and the news media when the NRA subsequently came to power. After the BBC interview, it was apparent that the Kampala government must allow further flights or risk a major media embarrassment. Woldridge was requested by BBC London to follow up the story with a visit to Kampala and asked me if he could come along on the next corridor of peace flight. I consented, hoping that further media coverage would guarantee the legitimacy of the operation. We flew back with the same charter plane and upon arrival in Entebbe, after clearing immigration and customs, loaded the plane with supplies from UNICEF's own small emergency store and filed a flight plan for Kasese. While the army security officer was nervous, he could not deny the local radio broadcast stating that President Okello took credit for granting permission for the Kasese flights. However, there was another snag, taking Woldridge along on the flight. He had been black listed by the previous Obote government and had only recently been accredited to work as a journalist in Uganda by the Ministry of Information. But given the favourable media attention generated over the weekend, a quick phone call to the Ministry of Information in Kampala, cleared the matter up. We soared off to Kasese in high spirits knowing, thanks to the radio communication link established with the UNICEF/ICRC team on the ground, that the local NRA commander would be expecting our flight.

Aside from an airborne interview, the BBC was only able to report what Woldridge saw. The NRA refused to talk about anything except the relief flights -- the issue of child soldiers was resolutely off limits. When we got back to Kampala, everyone welcomed the corridor of peace initiative publicly and further flights proceeded smoothly right up to the fall of Kampala in late January 1986.

Throughout the period in which the corridor operated, negotiations with government and the NRA were centred around the principle of reaching children in war. Success was made possible only because of the airlift of supplies to the NRA which emanated from within Uganda. Public awareness and multiple partners as well as public sympathy and support enabled UNICEF and the ICRC to stand firm on reaching civilians, especially children, through the corridor. The issue of child soldiers was subsidiary to the success of the corridor but nonetheless important in the arena of child rights.

The NRA came to power in early 1986 after a four-day siege of Kampala in which 2,000 people were killed. Journalists flooded in. Child soldiers were everywhere in evidence, a focus for the press primed by the previous October's BBC interview.[3] The boys were hailed by the local and African press as young "liberators" while the western media dramatized their small size and tender age and worried about the long-term psychological impact of violence and use of guns.[4] *"Were they psychologically impaired by war?" "Could they even resolve conflict without recourse to the gun?" "Would they become dissatisfied after a few years and seize power?" "Could they be integrated back into village life?"* These were some of the questions raised by the media and Ugandans.[5,6]

Donors and diplomats as well as NGOs were also concerned about the child soldiers. But they did not have a working relationship with the new National Resistance Movement (NRM) government and therefore expressed concern through quiet diplomacy or off-the-record statements to the press, leaving UNICEF to take the lead on the issue of child soldiers. A good dialogue was sustained between the NRM and UNICEF. By March 1986, the children were withdrawn from front line duty in a war which was to sputter on for three more years in the north.[7,8]

While no researchers followed the youngsters after my departure from Uganda in July 1986, we know that many were released from the army within six months of the NRM coming to power. Most seem to have gone back to school and have faded into the fabric of society. Some of those who remained in the army have been given educational opportunities. From discussions with Ugandans that I have met and corresponded with, I conclude that only a minority have continued to carry guns and live within military barracks. By 1990, even the youngest child soldiers had grown up and even if they had remained in the NRA, they would no longer be easily identifiable.

Thus, the corridor of peace was an unprecedented success in establishing direct relief across the lines of civil war within Uganda. Although the issue of child soldiers fizzled out with the passage of time, the corridor probably contributed to the NRA's decision to remove them from front line duty, releasing those who wanted out and providing a free education for those who went back to school. What was lost was the opportunity to follow up on the child soldiers and evaluate the impact of war and violence on their lives.

Sudan 1986-1989

Operation Lifeline Sudan, which began in 1989, was much more spectacular in order of magnitude; also much more complex than the corridor of peace in Uganda. My role with the operation began in mid-1986, when we were transferred from Kampala to Khartoum. Sudan is a divided country, with the Arab Muslim north comprising 75 percent of the population and a Christian African south, the remaining quarter. But, more important, development has been given priority in the north as compared to the poorer and grossly underdeveloped south. Civil war exaggerated these differences, shattering the economy, disrupting administrative services and depriving the people of the south of even the most basic means of survival. By mid-1989, the democratically elected government was overthrown in a military coup. The south experienced widespread famine while the north was racked by socio-economic instability caused to a large extent by the war. Although these generalizations do not tell the whole story, civil war has been at the centre of the country's problems since the early 1980s.

Among the terrible consequences were the estimated 250,000 civilians who perished in 1988 alone. Young children, youth, their mothers and grandparents, all died as a result of war-induced famine (see chapter 1). Where were the international community, the NGOs, the UN? What about the Geneva Convention with its legal definitions of warfare and the neutrality of the ICRC to monitor prisoners of war and provide humanitarian relief? The ICRC had more difficulty in implementing relief efforts within the Sudan than in any other country caught in civil war. Prisoners of war and the civilian population have been largely inaccessible. No one had sustained access to the combat zones in southern Sudan until late 1988. Then the ICRC began providing relief assistance but only on a limited basis to both government-controlled garrison towns and the SPLA area and that after months of negotiation following the donor invitation to the ICRC to solve the problem of providing relief. Even before, in 1986, when the ICRC sent a medical delegation to Wau, it was stranded for weeks because neither the government nor the SPLA would allow it to fly a rescue mission let alone operate regular supply flights. Despite this initial experience, the ICRC persisted and eventually received official permission to work on both sides of the war.

Judging from the camps of war-displaced inside the Sudan, civilian victims were primarily the elderly, women and children. While refugees in Ethiopia were primarily boys and young men, southern Sudanese men were the first to join sides or migrate to areas of security and opportunity, even leaving the country as refugees. Women and children were the primary victims among the displaced, the enslaved or the thousands who died.

We arrived in Khartoum in August 1986, and at my first meeting with Prime Minister Sadiq el Mahdi in early September, I immediately raised the issue of opening a corridor of peace into the south. In principle, the Prime Minister agreed that UNICEF vaccines and oral rehydration salts should go to the south and I left his office feeling optimistic.

In subsequent meetings, held at roughly monthly intervals, I pressed for permission to open negotiations with the SPLA. The Prime Minister's response was measured and incremental, giving a little more each time. This was more marked after his meeting with UNICEF Executive Director James Grant in New York in the autumn of 1986 during which Mr. Grant also made a strong plea for permission for UNICEF to work in the SPLA-controlled areas.

I soon learned that controversial issues such as relief to the south required a wide consensus of political, civil administration and army agreement. The Prime Minister told me to meet the Minister of Defence. Once the Minister of Defence was reasonably convinced, he asked if the Ministry of Health agreed as well as encouraging me to meet each of the southern regional army commanders and provincial governors. The message was clear: I had to go far beyond the Prime

Minister if the corridor of peace concept was to be accepted. By early 1987, perseverance and persuasion resulted in the Prime Minister insisting that any campaign should start by immunizing the children in the garrison towns before attempting to reach the SPLA-controlled rural population. Because it was an ultimatum of sorts and because it was necessary to have Khartoum's full support, we moved quickly to strengthen the cold chain and establish an immunization programme in Juba, Malakal and Wau, the regional capitals of southern Sudan.

Médecins Sans Frontières Holland organized a medical team which went to Wau in March and a vaccination drive was launched in Juba. By late April, the Prime Minister was informed that the major garrison towns now had immunization services. He proceeded to review the 1986 immunization results for the country as a whole and then established the next milestone in the negotiation process: *"Until half the children in the north are vaccinated, UNICEF should not bother about the south."* I countered that the children in the south were Sudanese citizens and therefore equally deserving but the Prime Minister would not be budged and I went away disheartened. Nonetheless, I was even more determined because he had now committed himself to allow us to work in SPLA areas once the 50 percent mark was reached in the north.

The UNICEF-WHO supported Ministry of Health immunization programme made rapid progress and by September 1987, I was able to carry our EPI graphs to the Prime Minister showing that half of all new borns in northern Sudan had been immunized with at least one antigen. He phoned the Minister of Health in my presence for confirmation. The point was that vaccination services had actually reached half the infants, not that half of all infants were fully immunized.

Aware that increasing numbers of emaciated children were coming out of the south than even a year earlier, it was more important than ever to reach the SPLA areas. We had now met two of the Prime Minister's conditions. Seeing my tenacious mood, the Prime Minister sent me directly to the Minister of Defence to negotiate flights to the SPLA-controlled towns. Since I had already briefed the Minister on the outline of the programme, both he and the Prime Minister agreed to my meeting the SPLA in Addis in order to negotiate flights directly from Khartoum. After a series of 11 meetings with the SPLA, I realized that the proposed corridor of peace would not be easily accepted by the liberation movement which preferred a cross-border operation. *"We want your relief and agree to UNICEF working in the south. We appreciate the water programme which has benefited our people in the 1970s and early 1980s, but we prefer that you turn your aid over to the Sudanese Relief and Rehabilitation Agency* (humanitarian wing of the SPLM). *We will take the relief ourselves into the south from either Ethiopia or Kenya."*

After numerous discussions with the SPLM, they took the predictable position that we should work from Kenya on a cross-border basis, full stop. They reminded me that the barges used to carry World Food Programme (WFP) food from Kosti in the north to the garrison town of Malakal in early 1986 had arrived loaded with military supplies and soldiers.

This made my task much harder; but eventually, after meeting with members of the SPLA high command, an agreement was reached to allow the UNICEF Twin Otter to fly to three northern-controlled garrison towns and to three SPLA locations. At a meeting in early November, it was agreed that flights originating from Khartoum would begin flying the first week of December and thereafter the first week of each month.

Meetings with the Minister of Defence continued; however, he initially rejected the idea as militarily unacceptable even though we had discussed the principle at length throughout the preceding year. Once the southern army commanders agreed, the Khartoum hierarchy started to give us wary consideration. Finally the Prime Minister called the Minister of Defence and me to a meeting during which time I reported that the Minister of Defence had not, as yet, agreed to a flight schedule or delivered the clearance to fly from Khartoum to SPLA-controlled areas. Shortly thereafter the Minister consented, provided the southern commanders would again agree.

Flight permission was arranged through an army colonel assigned to our flight operations in the Ministry of Defence army headquarters. Similarly, an "implied consent" system was worked out with the SPLA in which we informed the SPLA four days prior to all our flights to the south through UNICEF Addis Ababa. This system allowed all flights to proceed unless the SPLA explicitly advised us against a specific flight on a specific day. Two of the three southern army commanders finally agreed that the corridor of peace would be preferable to clandestine cross-border flights.

A one-and-a-half page army order was drafted and presented to the Minister of Defence for signature in late November. It was never to be approved. On the day I was to receive the written permission, all hell broke loose in Khartoum with the mass mobilization of the army. The SPLA had successfully attacked and captured the northern town of Kurumuk near the eastern border with Ethiopia. At what was to have been my final meeting with the Minister of Defence for the final go ahead for the corridor of peace, the Minister instead waved the order in front of my face and rejected it in its entirety. I argued that this was the nature of war and that one setback should not be allowed to jeopardize months of negotiations for the children. Further, I maintained that the northern army would gain from the positive publicity if the corridor of peace were opened. But no amount of pleading could change the Minister's mind. The principle and months of preparatory work were lost.

This severe set back, at least on the government side, could not be allowed to sour relations with the SPLA. While meeting our children who were transitting through Addis en route home from boarding school in India, I had my only unannounced meeting with the SPLA. Upon reaching Addis, the SPLA restated their desire for the immunization programme to be carried out across the border. We parted on friendly terms.

In mid-January 1988, I flew with the UNICEF-owned Twin Otter to Wau to evacuate the MSF Holland medical team. A subsequent account of the nine months of horror there in which three massacres of civilians took place appeared in an anonymous paper "Sudan's Secret Slaughter." [10]

With this and other reports, the NGO and donor community became increasingly alarmed by the evidence of famine, the non-cooperation of the Sudanese authorities and the press stories of government atrocities. Donors also felt frustrated at their own failure to deliver food aid to the south and unanimously agreed during an aid coordination meeting in Khartoum to sponsor the ICRC as the best means of reaching the famine victims in the south. Meanwhile, my monthly meetings with the Prime Minister had resumed and he had shown himself sympathetic to the contention that the Kurumuk setback should not derail the corridor of peace. Between January and March, 1988, the UNICEF aircraft had completed nine flights south including trips to Malakal, Wau and Raja. UNICEF was now flying to all southern capitals as well as to Yei and Yambio while other aircraft belonging to UNDP and NGOs were restricted to northern destinations and Juba in the south. Nile Safaris, a private charter company, was also allowed to fly to the south. The SPLA continued to give their implied consent to these flights south even though we were not flying to SPLA locations.

There was an increase in critical press coverage regarding the government's unwillingness to provide relief to the south, including frequent reminders that the major donors had put all their resources behind the ill-fated Operation Rainbow mounted in October 1986. [11] This WFP operation, supported by 11 major donors, not only failed but resulted in the expulsion of the head of UNDP from the country amid worldwide press coverage. UNICEF simply could not afford to risk a similar experience. Four major NGOs which had been active in the south were expelled from Sudan in late 1987. The government refused to give any reason for the expulsion. Separate UNICEF flights to the south were thought by some donors and NGOs to potentially compromise ICRC negotiations. *"For much of 1988, the US gave top priority to supporting ICRC efforts to negotiate an agreement with the protagonists, at the same time restraining UNICEF's wish to assist in SPLA areas"* (p. 115 of *Humanitarianism Under Siege*, Red Sea Press, 1991).

Tensions were mounting between the NGOs and government over human rights violations in the south. Obstacles placed in the path of NGO relief efforts in the south and to displaced southerners in the north by the government resulted in a concomitant tension between the NGOs and bilateral donors. The NGOs became increasingly outspoken, especially to their media contacts and European politicians, calling on their governments to place conditionality on official bilateral aid in an effort to force government policy changes. UNICEF received some "fall out" from discontented NGOs.

In this charged and frustrating atmosphere, there was a very real possibility that I might be declared *persona non grata*. The head of UNDP, serving also as special representative of the Secretary-General for relief operations with the rank of Assistant Secretary-General, had been expelled in late 1986 over his "unauthorized" discussions with the SPLA surrounding the ill-fated Operation Rainbow. Nonetheless, UNICEF headquarters decided to mount a limited cross-border operation from Kenya while respecting the ICRC and donors pursuits in Khartoum. This was implemented with minimal involvement of the Khartoum office.

Cross border

MSF Holland was our principal partner in the cross-border operation along with two other NGOs as well as Norwegian People's Aid. UNICEF support was channelled to the NGOs from the regional office in Nairobi. With limited success, came media interest. The newspaper reports raised eyebrows among the major donors in Khartoum who once again voiced their concern that UNICEF was out of step with the ICRC initiative. The ICRC had not managed to provide relief to the south up to September 1988 despite increasingly graphic press reports of famine and starvation except on a limited basis through an emergency hospital at Narus just across the Kenya border. Reports circulated in Nairobi and Khartoum that the government was aware of UNICEF's involvement in the cross-border operation and was preparing to take UNICEF, and perhaps the entire UN system, to task for it.

The international news media focused anew on the Sudan when unprecedented floods, the worst in the century, flooded Khartoum in August 1988. The Nile burst its banks and Khartoum was deluged by 24 hours of heavy rain. As the flood water receded, camera men and reporters, guided by NGO workers, concentrated world attention on the plight of displaced southerners living in appalling conditions in make-shift camps around Khartoum. The reporters followed the displaced back to their camps at Muglad, Babanusa and El Meram and eventually to Abeyei in southern Kordofan. A massive famine was everywhere in evidence and was captured by the BBC and CNN and projected into homes around the world in the autumn of 1988. With each newspaper report and broadcast came increased speculation in Khartoum that the government was upset and would crack down on the cross-border operation. By the third week of September, I was summoned to a meeting with the Minister of Foreign Affairs and the Prime Minister. At Foreign Affairs, I received a firm yet understanding hearing: *"Yes, the government is upset... but understands the motivation of UNICEF to be humanitarian."* However, my reception at the Prime Minister's office was ice cold. He acknowledged neither the months of discussions nor the agreement with the principle that UNICEF had a mandate to meet the needs of all Sudanese children, be they northern or southern. *"Your well-intentioned cross-border assistance is helping the SPLA who are the enemy of democracy. Thereby, you are prolonging the war and contributing to the death of my soldiers,"* he said. This statement was later quoted in an article in the *New York Times*, which eventually published an editorial on the subject. [12]

Children were dying and from a UNICEF perspective we had to press ahead. Executive Director James Grant met the Foreign Minister in New York in early October, and while accepting that UNICEF should curtail the cross-border operation, Mr. Grant pressed for, and received, a verbal pledge that we could work with the ICRC in the south. This agreement to curtail cross-border operations was again picked up by the media, which appeared that we could do nothing right and whatever we did do was criticized from one quarter or another.

Operation Lifeline

Two significant developments occurred in the autumn of 1988 and early 1989. First, the international media coverage of the famine exposed the position of the government towards humanitarian relief to the south. [14] This increased political interest in western donor capitals and, importantly, among politicians and resulted in a series of political visits, including two by United States Congressional groups and a host of European delegations. Country after country began declaring their intention to provide relief assistance to the south as a humanitarian priority beyond Sudan's claim to sovereignty.

Second, the ICRC received the "go ahead" to work on both sides in late 1988 after nearly a year of stop-start discussions on modalities. A limited cross-border operation based in northern Kenya was expanded and began providing assistance to three locations held by the SPLA. Simultaneously, flights commenced to Wau and Malakal, to assist the garrison towns as the *quid pro quo* for the north from Khartoum.

When the log jam broke, the UNICEF office in Nairobi requested a reappraisal of the decision to halt cross-border assistance. This followed a fresh request from the SPLA for a UNICEF vaccination programme, something that had been discussed but never implemented.

By early December, it was clear that three specific corner stones needed to be put into place to enable UNICEF and others to build a programme of assistance to the south. I set about doing these in direct consultation with the Executive Director, James Grant.

The first was to get a letter from the Prime Minister explicitly approving UNICEF as a partner in the ICRC's southern cross-border operation. Since Mr. Grant had already obtained the Foreign Minister's verbal consent, it was logical that I should follow up. Some Khartoum-based donors did ask "Why UNICEF?" The remaining cornerstones were to obtain agreement from the SPLA for UNICEF to work with the ICRC and to negotiate an agreement with the ICRC.

Soon UNICEF was in the air again, this time on a flying visit to the SPLA-controlled area inside southern Sudan. One afternoon was all that was needed to reach an agreement, as the SPLA had encouraged us to step up our assistance for many months. Negotiations with the ICRC then began and the basic relationship was concluded for subsequent signature by the ICRC and UNICEF in Geneva. Only the Prime Minister's written approval was now required for UNICEF to work with the ICRC on a cross-border basis. The Prime Minister agreed that I could pick up the letter up from his personal assistant. But the appointed time came and went and no letter until finally I asked the outgoing World Bank Representative, Jasdip Singh, to raise the issue regarding the letter during his farewell visit to the Prime Minister. He came away with the letter signed.

In early February 1989, the Secretary-General appointed James Grant as the UN Coordinator for Relief Operations in the Sudan. A high-level donor-government conference was held in Khartoum which approved Operation Lifeline. Massive relief deliveries were planned for April to send food to the south prior to the onset of the seasonal rains which isolated the countryside during the summer months. Food relief targets were met, surpassing all previous years, despite the difficult schedule. Food deliveries from Khartoum were particularly problematic. However, for the first time, vaccinations were given to southern children in the SPLA areas. The book *Humanitarianism Under Siege* provides a full account of Operation Lifeline Sudan, which ranks as one of the most difficult and massive relief operations ever mounted.

Discussion

If children grow up knowing only deprivation and famine, will they be socialized into a world of violence and despair? A whole generation of Sudanese children are growing up illiterate: schooled only in one of the rawest civil wars the African continent has seen. How will this entrenched civil war end? What UNICEF did to reach children caught in war, was only the start of a long and difficult course. This is the story of what I was able to do with the help, involvement and support of numerous others who took initiatives which led to risks of their own. This is especially true of the ICRC and the NGOs working in the south, and also the media professionals and donors who travelled in the Sudan. Our brief successes for children are only a beginning. A new world ethic which rejects unnecessary political hindrances to provide children with their rights has to be encouraged and supported. What about the continuation of Operation Lifeline? The mandate of the Convention on the Rights of the Child is a giant step forward, but now we must transform children's rights into our social obligation.

We need and must generate a sustained global advocacy campaign to give children a chance. In my work in war situations, I have never met anyone, from front line army commander to guerrilla fighter, from Prime Minister to opposition leader, who has not agreed that we should all work to leave the world a better place for our children.

References

1. Wiebe, P.D., and Dodge, C.P.(eds), *Beyond Crisis: Development Issues in Uganda*. Makerere Institute of Social Research and African Studies Association, Kampala, 1987.
2. Dodge, C.P., "*Corridors of Peace in Uganda,*" *Assignment Children*, Vol. 69, No. 72, UNICEF, New York. 1986
3. Harden, B., and Hilsum, L., "These Young Fellows Laugh at a Dead Body," *The Washington Post*, early 1986.
4. Legum, C., "Uganda's Army of Child Soldiers," *Third World Reports*, London, 13 February, 1987.
5. Hilsum, L., "Not Too Small to Kill," *Children First*, London, Autumn, 1986.
6. Amin, M., "Children of Terror," *The Nation Sunday Magazine*, New Jersey, March 10, 1986.
7. Gargan, E.A., "A Child's Lot in Uganda: At 14 a Combat Veteran," *The New York Times*, New York, August 4, 1986.
8. Harden, B., "Political Violence Has Left Psychic Scars on Ugandan Children," *The Washington Post*, Washington D.C., July 15, 1986.
9. Bonner, R., "Famine," *The New Yorker*, New York, March 13, 1989.
10. Anon., "Sudan's Secret Slaughter," *Cultural Survival Quarterly*, Vol. 12, No. 2., Cambridge, 1988.
11. Minear, L., et.al., *Humanitarianism Under Siege*, Red Sea Press, New Jersey, 1991.
12. Editorial, "A Killing Silence in Sudan," *The New York Times*, New York, 17 January, 1989.
13. Perlez, J., "U.S. Relief Effort Blocked in Sudan," *The New York Times*, New York, October, 1988.
14. Hendawi, H., "The Vulnerable in Sudan are all Dead," Reuters in The Daily Nation Nairobi, 13 December, 1988.

CHAPTER 7

HELPING THE CHILD

Magne Raundalen and Atle Dyregrov

The cultural dimension

Traditional cultural and societal values may provide the basis for providing practical help war-traumatized children. Child rearing and socialization practices should be explored and utilized in efforts to provide therapy. Two important cultural dimensions have a role to play. The first is child culture, which consists of songs, dances, rituals, rhymes, storytelling, fairy tales, dramatic play, creative activities, sports, games and social interaction. Such activities have the potential to overcome after-affects of stress and trauma by allowing children to express their feelings in the familiar context of their own routines and traditional culture.

Second is adult culture, which through explanations, religious rituals, interaction between the child and parents, guardians and/or caregivers can yield the means to help traumatized children.

In Mozambique, we identified a reluctance on the part of older teachers to help war-traumatized children because they thought it best that children not discuss traumatic events. In their view, silence concerning man-made tragedy was the best cure. The younger teachers, however, were more receptive to encouraging children to express themselves.

Aggression and war

Normally, in peaceful times, when the necessities of life are secure, aggression and violence are not needed to survive. Nonetheless, most children acquire "survival" traits. Culture defines sex-roles for us and although these roles may differ from tribe to tribe and country to country, generally males are encouraged to have more aggressive roles. Traditional values of all cultures, while encouraging some forms of aggression, provide constraints on the use of aggression and violence. Parents, relatives, teachers and other adults set the limits by scolding, reprimanding and punishment.

We know, from the study of child abuse, that there is truth in the adage "violence breeds violence." Parents who have been abused themselves, are much more likely to abuse their own children.[1] More relevant are such findings as the fact that children brought up in Northern Ireland who have been inculcated from a very young age to hate their enemy fulfil the prophecy that the cycle of violence is passed from parent to child.[2] Yet, even allowing for the vast differences between Northern Ireland and Uganda, our own research during Uganda's civil war contradicts this simple conclusion. We originally thought the violent years of Amin and Obote would create a generation of violent children, but, although the children showed anxiety, depression and grief, we found almost none harbouring revenge motives. Furthermore, when we interviewed and asked the children what they wanted to do in the future, they gave positive answers. Despite having lost a parent, the majority had healthy aspirations, wanting to be a nurse, a doctor, a driver of the Red Cross, an employee with UNICEF, and so on. They were not preoccupied with thoughts and acts of hate and revenge, although some expressed negative feelings. Most had enough of the armed men and killing and simply expressed their hope for peace.[3]

On the other hand, there were many trained child soldiers in Uganda who had helped the NRA in their guerilla war. We do not know if these children were motivated by a desire for vengeance when joining the NRA, but it is clear that they became good" soldiers and performed all kinds of violent acts. The NRA explained the children's training in the use of modern weapons as an outgrowth of traditional cultural values which encouraged young boys to use sticks and spears to defend livestock from animal attacks (see chapter 4).

Civilians in the Ugandan countryside, particularly the Luwero Triangle, believed to be sympathetic to the guerrillas, were frequently harassed and killed by government forces. These victims of war, many children, exhibited after-affects characterized by anxiety, Post Traumatic Stress Disorders, depression and grief. Despite this, the children wanted to participate in peaceful and even "healing" activities for their troubled country and had a healthy future perspective. While these positive attitudes seemed real enough, we fear that the optimism is a thin veil covering underlying bitterness and may be even aggression. These victims could shift their emotions to aggression and violence if a peaceful and meaningful future is not sustained.[5] Psychological research confirms that depression and grief is sometimes relieved by aggression and violence if no other solution is available.[4,6,7]

Summary aggression

- Experiences of war in the home environment resulting in destruction of buildings, flight from home, and, loss of parents and siblings can cause grief, anxiety and depression. However, vengeful attitudes do not necessarily surface among these children. We are aware, however, that traumatic events impact on their future prospects and can generate aggression and violence later if physical and psychological recovery does not occur.

Panel 7

"Now they fear if someone bangs the door."

Living in Uganda during the war, Rose, 36, married with five children ages 4 to 17 wrote:

"I was born in Fort portal and schooled there until I married on the last day of 1966. My husband was working in the Ministry of Education so we moved to Kampala together. I spent most of my childhood in the palace because my father was personal guard to the King. I was a favorite of the King and he used to pick me up and play with me in my happy childhood at the Kingdom of Toro.

"In 1973, we were recalled by Amin. When my husband went to the Ministry of Foreign Affairs, he was told by friends to keep away from the ministry as it wouldn't be safe. He would be thought of taking the position of someone else. He started going to the Ministry of Education. We were told the same story. Through friends we were able to find an empty house to stay in as many Europeans had run away. My husband started to work at Makerere University and was threatened very often.

"Anyone that knocked at the door, any car that came, we would be frightened. We drew the children maps to relative's places in case we weren't there when they got home. Telling them who was a real uncle and who wasn't. There was a commission to investigate the running of the university and it sounded like the personnel officer (the narrator's husband) was responsible for mismanagement. He wouldn't allow just anyone to get a job and insisted for the right papers. As a result he was unpopular with the president's office. We planned that he run away and I take the children to Fort portal, wait to hear from him and then join him. Then the war came in 1979. We stayed in our home until 10 April 1979 although we were in danger as the neighbour's house had been hit. They were not killed at this time but three weeks after the war, murderers came in and killed the mother and her boy of 15 at which time we heard terrible screams. We went to the residence hall and stayed in the basement with six other families. We cooked upstairs and shared food. There was even a baby one day old.

"We tried to keep things away from our children. But the bigger they got, our friends got in trouble and they would talk about bombing and looting. Thieves came into our house in 1983 and took practically everything we had. They came in six different times. We never had guns. They simply emptied the rooms. So for one whole year, the older children slept in one room. If lights went out, we wouldn't eat but hide under the blankets. Now they fear if someone bangs at the door. Still they don't go into a dark room alone. The children don't talk to strangers for they fear soldiers and thieves. They turn the other direction when they see a soldier. Our daughter hardly ever goes out to the main gate."

Rose. Kampala, Uganda, 1986.

- During war, children are sometimes socialized to violent behaviour: as when a child sees a father, uncle, or older brother preparing to fight or commit acts of sabotage; or when the child listens to adults degrading and blaming the enemy for their troubles. Destruction of the enemy, including the killing of their wives and children, may become not only acceptable but may even be rewarded in the skewed social context of war. Children may absorb a cultural indoctrination of hate and violence, often based on religious or ethnic supremacy, which enables them to fight and, in some cases, commit atrocities.
- Historical and cultural values may change over time, allowing different levels of accepted aggression and violence in a society; hence the varying degree to which children are socialized to violence. This may ultimately lead to a situation where children are given opportunities to participate in violence. Aggressive behaviour may escalate in the classroom and through interaction with peers as a result of war. Lawlessness is frequently a by-product of these cultural changes where traditional values are severely eroded. (For a discussion of the breakdown of traditional limits on violence see Denis Pain in *Beyond Crisis*: Development Issues in Uganda.)[5]
- Children may get politically involved in extreme situations where there is no solution in sight. A feeling of despair and loss of future prospects, in combination with passive parents, may express itself in physical force, as has been observed in South Africa and the West Bank. Since Soweto in 1976, the children of South Africa have been an important part of the liberation fight. The same applies to the West Bank since the *Intifadah* in November 1987 when children took on a major role in fighting the Israeli soldiers. [6,7]
- During times of war, aggression and violence are legitimized by norms of behaviour previously unacceptable. Unfortunately, the aggression sanctioned during war, all too often persists after the conflict ends or spills over from the battlefield into the arena of normal social interaction. Violence may become accepted as a legitimate reaction to stress and frustration. War time aggressions may affect the upbringing and socialization of children which may cause societal problems long after the fighting has ceased.
- Many children, forced to flee their village home because of war, are also affected, even though they did not have direct contact with armed forces. These include refugees, the internally displaced and street boys, whose lives are both difficult and dangerous and who often practice aggressive anti-social behaviour.

The Convention on the Rights of the Child recognizes the need for psychological care. We suggest that psychological care be integrated into relief and rehabilitation programmes in much the same way as immunization is provided for as a part of medical relief assistance.

Children's preparedness for the future

As children grow and develop, they combine all relevant aspects of their life: parents, family, their own capacities, the national and, eventually, the world situation, and composite these to form a calculated future plan which is linked to their developmental stage.

Their future prospects may be related to the capacity of the family, the clan and the nation. The events of the region and the foreign policy of other countries, such as Mozambique's relationship to South Africa, will become important for any child in Mozambique. In the broadest scope, they calculate parameters from global problems to their own preparedness for the future. "Nuclear winter" after a nuclear war or the green house effect and the destruction of the ozone layer are considerations which influence the future prospect for children who read newspapers or watch television regularly.

But how does this relate to our goal to reach children in circumstances of war? Because the children are at varying stages of their development, we must respond differently.

For the first three years of life, the child mostly relies on the mother, the family and relatives to provide daily needs. During this age of dependency, children's relationship to war, famine, and political unrest are secondary to the security which mother and the immediate family provide. Times of war may affect children's basic trust in even their own mother if she is unable to provide basic nurturing.

What can we do during this dependency stage to strengthen preparedness for the future? In order to reach the child, we must first approach their parents. On a national level, the media, especially radio, may be the only practical means to communicate with a significant proportion of mothers and fathers who have very young children. Religious services such as prayers at the mosque or church services provide important communication channels. Immunization sessions are another possibility. Nervous parents can adversely influence small children. If the mother is upset and cries, the parents should be encouraged to talk to their young child about why she cried, and why they are anxious. In cases of forced separation, parents should try to make arrangements with guardians known to their children, or, better still, to plan, if it is possible and not too dangerous, to keep their children with them. It is important for children who belong to a neighbourhood play group, pre-school or kindergarten to keep their daily routine even in times of war. Their feeling of security affects their future prospects.

Any dramatic event that may harm family members, will affect the child and his or her expectations for the future. Many children are old enough, by age seven, to have incorporated into their future, concepts of illness, accidents and killings. This is especially true if events have taken place in the neighbourhood and affected their family.

As children grow older they absorb the opinions and views of adults with whom they interact. Their own prospects for the future incorporate opinions and views of acquired information about cause and affects of empirical events. Small children during the first years of school, can be taught to prevent some of the negative prospects they observe. But they will feel powerless when confronted with events that have negative impact on their lives such as an unemployed father, price rises, lack of food security in their own home or the occurrence of war, desertification and other macro-level events. How can we strengthen their prospects for the future in times of war? How does a traumatic event effect a child's belief in a safe future? How can we help children living in an unsafe environment to keep up hope? These are vital questions to which we constantly seek answers.

Parents and teachers should provide coherent explanations of current events so that children can develop an understanding which allows them a predictable future. This can be done by explaining the causes of war and violence, not simply as the events occur, but through developing their understanding of social and political problems. Parents should not attempt to keep information away from children on the assumption that they are too young or too immature to understand - rather they should explain the situation to their children in as simple and straight forward a manner as is possible while reassuring them that the family will cope with the situation.

Reaching children through teachers

One of the most recognized ways of helping the war-traumatized child is through manuals for teachers. [8-12]

In most African countries, despite difficulties of large classes and limited training in child psychology, teachers remain the best avenue to reach war-traumatized children and have the following advantages:

- They are familiar with the local culture and have insight into the norms, rules and customs.
- They know the national laws and the rights and obligations.
- They often know the individual child and are able to register personality changes.
- They are trained in transferring knowledge to children.
- They are well known and are generally trusted by the community.

- They and the school represent stability and often provide a secure base in times of war.
- They may know the parents or the guardians of the child and are respected as advisors.

Expression

It is of the greatest importance that the teacher adjusts expressive therapy to the cultural setting and background of the children. A wide range of techniques are available ranging from group interaction to writing poetry. Creative material like clay can be used to re-create events of war. Dance, singing, acting, participation in sports and other forms of physical exercise are also suitable forms of expression, provided they are encouraged within a culturally acceptable manner.

Understanding

Teachers may use the principles of insight therapy to help children understand why they react as they do. To help the child grasp his own reaction patterns through psychological explanations of how anxiety, for example, is conditioned and why sadness follows loss and may transcend to aggression. An important aspect of this cognitive coping is the understanding of the situation causing stress and trauma.

Training

Behaviour therapy training can range from systematic desensitization to helping the child through a visit to a place where he or she was frightened. Some behaviour is abnormal because it occurs too often such as hyperactivity, anxiety and aggression. Other actions occur too seldom such as retarded speech, helplessness or the failure to learn in school. Abnormal actions also include wetting the bed or crying during school lessons because they occur at the wrong time. Behaviour therapy can be used to treat fear and phobias that have been acquired during times of war. Two methods are highly affective in treating children's fears. The first is called direct conditioning and consists of providing a stimulus that arouses positive responses simultaneously with a gradual presentation of the feared stimulus. The other is social imitation. This means showing the fearful child that a non-fearful child is interacting with the feared object.

The direct conditioning procedure has been developed into an important treatment technique called systematic desensitization: training the child to react less intensively to the feared stimuli.

Gina, Gaza, Mozambique

Gina, who was interviewed four times in Maputo, suffered nightmares since her home was attacked. Gina watched her grandfather being killed and saw her aunt beaten and die in the house which had been set ablaze. Gina was kidnapped two times by the bandits. On one occasion she was taken together with her mother, another time she was forced to walk for three days to the bandit's base. Along the way, people were killed; a mother with a baby who tried to hide under a car was shot and Gina watched the bandits cut the throat of another person. A woman who refused to obey orders was killed.

First of all, Gina needs to talk about what happened with her mother so they can share some of the common horror they experienced, particularly if nightmares and thoughts reoccur. If it is not possible for Gina to talk to her mother, the story could be shared with her teacher who should encourage Gina to express as many details as possible about the painful memories, including the loss of her close relatives. Gina should be given the opportunity to show her feelings, despite the pain, by acting out some of the situations, to cry and to express aggression towards the bandit's brutality. It may be helpful if the teacher writes down Gina's story as the child talks or for Gina to write it down and to encourage Gina to draw some of the more appalling scenes she remembers. The drawings and story should be kept for future reference.

Gina should be encouraged to talk and work through the atrocities she experienced several times. Particularly the incident when she was kidnapped alone and how she managed to escape and walk for three days without food, and about the killing of her relatives.

Joao, Manica, Mozambique

"*I will never forget that warm Wednesday afternoon when I met the terror of the armed bandits myself. I and my brother were driving along the main road together with a man who had asked for a lift. They blocked the road and we had to stop. One was armed and I had no time to observe the other person. They asked the person who was travelling with us: 'Do you want to die standing upright or lying on your back?' The man said he was the priest in the church near by, and he therefore wanted to die on his knees. Then they shot him. I looked for my brother, but he was already dead. Blood was colouring the new white shirt he was so proud of. Then one of them said: 'What shall we do with that little bastard?' Then the other one said: 'Lets go, we do not waste a bullet on him!' Before they went away they set fire to the car.*

"*Can you tell me, my brothers, what is the solution to this problem? If it has been a conflict between races, I could have understood them, but those who killed my brother and the priest were Mozambicans like me!*"

The text we have quoted is a composition Joao wrote at school about "*war and violence in my life.*" The teachers who gave their pupils this composition, explained that many of the children became very upset while writing. Later, the children wanted their teachers to listen to their stories as they remembered more details.

One way to help children like Joao cope is to let them write down their painful memories and afterwards to talk about them in the classroom. The teacher has the opportunity to encourage them to share and thus make it easier for them to recover. Furthermore the teacher has the opportunity to explain the political background for these unbelievable events and therefore to help eliminate the feeling that they live in an unpredictable world.

Carlota, Maputo, Mozambique

Carlota's story is quite similar to the two previous ones. The terror starts at an evening dance at school.

"We were singing and dancing and then suddenly the bandits came. They said they were from the other village and started dancing too. Suddenly they turned off the music and we tried to leave when they started shooting. I ran away and hid in an old house. While I was hiding, the bandits discovered me. They forced me to go with them and when we passed through my village, they put fire to my grandmother's house, but it was empty. They stole food and kidnapped a lady. We were a large group and they killed so many that day. They asked a woman to carry a bag of salt. She carried it halfway but then she said, 'I am so tired!'.

"Oh, you are so tired?' the bandits said.

"Yes, I am so tired,' she replied wearily.

"OK, you will rest, forever,' they said.

Then they killed that woman and she died right in front of us."

Carlota managed to escape together with two other girls after two days and nights at the bandits' base.

Carlota talked to many people about her experience and found it particularly helpful to talk to her mother and teacher who visited their home. She said that she felt better when she told us what had happened. However, she was frightened to attend school when a drama or dance was held. The teacher and her mother explained to her that this anxiety appeared because it reminded her of the day she was kidnapped and assured Carlota that kidnapping would not take place in the city of Maputo where they had moved. This assurance worked for a few days, but then the anxiety reappeared, so strongly that Carlota hardly dared to walk outside.

Carlota experienced conditioned anxiety and her anxiety generalized to situations which reminded her of the evening she was kidnapped. This anxiety did not disappear only by soothing talks and good explanations, Carlota needed to be involved in a training program consisting of a step by step approach to the schoolyard meanwhile receiving positive reinforcement.

At first Carlota's mother was hesitant to accompany her daughter to school, or to stay there for few hours and collect her when the school day was over as this schedule interfered with her numerous daily tasks. But on the third day, one of Carlota's classmates replaced the mother. Carlota's anxiety did not bother Carlota to the degree that it prevented her from going to school and by the end of the week, Carlota felt quite comfortable when she went to school and returned together with her classmates.

Carlota will still live with reactions and after-affects to the terror she experienced when she was kidnapped, but now she is at least secure enough to leave the safe base of her home and to stay at school.

Eduardo, Chimoio, Mozambique

Eduardo was one of the best pupils at his school. He could read and write and was very good at mathematics and had taught himself English by listening to the BBC and reading old newspapers he found. Eduardo and his family had moved earlier from Tete to the Manica Province because of destabilizing activities and direct attacks from the armed bandits. From the time he was eight years old, Eduardo experienced many threats and much anxiety concerning family members until he absolutely refused to go to school after an attack by the armed bandits. One of his school friends disappeared that day, and the assumption was that he had been kidnapped by the bandits. Eduardo managed to escape before the school was blown up, but he heard bombs and shots while he was running. Some of the other pupils were injured but no one was killed.

For three weeks after the attack, Eduardo did not talk at all although he understood what his parents said. He started talking again when he was alone with the cattle and gradually resumed normal speech; but nobody talked to him about what had happened, because they believed that if they did, he would stop talking again. This is a common but wrong-headed way of reasoning. The best way to help Eduardo would have been to have the teacher visit his home and talk about school and how the other pupils missed him. He should have been encouraged to read his books and study his lessons. Every week the teacher should have visited the home and talked about what happened when Eduardo and his family had to flee, about the bombing, and about his friends and the daily events at school. After two or three weeks, Eduardo should have been encouraged to accompany the teacher to school just for a visit until he was willing to attend classes. If this approach didn't work, Eduardo could have been introduced to another school in the area, preferably one that had not been bombed and with no associations with killing and kidnapping.

Two possible psychological phenomena should be mentioned that might explain Eduardo's reaction in part. The first concept is called survival guilt, which, since one of his best friends was kidnapped, Eduardo may have suffered from because he escaped and did not help his friend. Eduardo should have been assured that he did the right thing, that no one could have stood up to the bandits and that all the others fled. He should be allowed to cry or express his grief over losing a class-mate, to share his experiences with the other students during a special session after his return to classes.

The second concept deals with intrusive images or thoughts. Many children experience horror in an extreme situation like the one Eduardo fled from and this was probably the cause for the loss of his speech. Events that have left painful images such as a soldier threatening him with his gun, the screaming of an injured classmate, the wall down by a bomb or adults crying needed to be talked about.

Nelson, Nampula, Mozambique

For two years, Nelson lived in the streets of Maputo. He left the village of Nampula by sneaking into an airplane shortly after bandits attacked his home. He knows that his mother and four siblings are still alive, but the mere thought of going back to his village sends waves of anxiety through his body and tears come to his eyes. It fills him with tremendous fear thinking about what happened when his grandmother was killed and mutilated, when people were left dying on the roadside, and when guns and bombs exploded. He felt he would be the next to be hit.

One can guess that Nelson had other problems, in addition to his fear of returning to his village. Nelson's father ran away and another male replaced him; probably providing a model for Nelson's escape; however, Nelson still feared returning to his village. When talking with Nelson outside the big shops of Maputo, he expressed concern for his future saying that he would be destroyed by the harsh and aggressive street life. He said that living on the streets was no human life at all and he called himself a lost person. Nelson seemed to be an intelligent boy who learned to read and write and speak English quite well.

In Nelson's case, all avenues for the reunification with his family should have been explored. As he was street wise, and knew very well from people he met on the street, from newspapers and the radio, that attacks still occurred in his part of the province, he should have been given the opportunity to talk and work through his anxiety. Ideally, he should have been provided with a foster home and a chance to return to school.

Therezinha, Inhambane, Mozambique

Therezinha's family fled the village of Inhambane after several attacks by bandits. No one was hurt, but the bandits entered the house, forced her father to give them money and food. They fired shots in the air when they left.

Ever since, loud noises, shouting and scolding have made Therezinha upset and she bursts into tears. Yelled at, she responds with hysterical seizures and takes hours to quiet down. Therezinha became more and more withdrawn and finally could only whisper her answers when the teacher asked questions in the classroom.

Therezinha's hysterical reaction to loud voices was not caused by the sound, but by the aggressive content causing avoidance behaviour. Even gentle corrections were intolerable for Therezinha, because they elicited associations with the violent bandits.

Therezinha should have been encouraged to express her fear of aggressive people. She should have been allowed to show anger towards the bandits and reassured with accounts of what the government was doing to curb armed bandits. Carefully and gradually she should have been given insight and understanding to why she reacted the way she did; with training to tolerate more and more noise as long as it was not directed towards her. She should have been rewarded for gradually raising her voice when speaking.

Akmed, Juba, Sudan

Akmed lived on the streets of the capital Khartoum. We met him at a street boy's centre, where he was able to get a shower, some food and was offered reading and writing lessons. Although Akmed had schooling, he failed his classes and continued to forget all he had been taught. Trying to fill in his psychobiography had given us rather meagre results, but we knew that he had fled a war zone where 40 percent of the children had lost one of their parents in civil war.

Akmed often went into the office of the social worker. He did not say much, but showed her a finger where he needed a band-aid or pointed to his stomach indicating pain. He also complained of severe headaches and, when asked, he admitted that he was sniffing petrol. Akmed had all the signs of depression; a slow and blurred way of talking, was extremely tired in the morning, and was always worried about the well being of his family in Juba.

After he started talking, we learned that he had lost two brothers, one because of illness, the other killed by gun shot. His father had joined the guerrillas and his mother had married again to a person who rejected Akmed. Akmed had been encouraged by his uncle to leave for Khartoum and got permission to sit on the back of a lorry together with some other boys. During the trip there had been fighting about food and belongings, and one of the boys fell off the truck under the wheel where he was killed. These memories filled Akmed with horror and guilt which reactivated the "survival guilt" he felt when his brothers died.

When Akmed was able to express all this to the social worker, his depression left him and he joined Street Kids Incorporated, delivering messages by bicycle in Khartoum as one of their most reliable couriers.

References

1. Dyregrov, A., Raundalen, M., "Children and the Stresses of War: A Review of the Literature," in Dodge, C.P., Raundalen, M. (eds.), *War, Violence and Children in Uganda*, Norwegian University Press, Oslo, 1987.
2. Harbison, J., *A society under stress: Children and young people in Northern Ireland*. Open Books, London, 1980.
3. Dodge, C.P., Raundalen, M. (eds.), *War, Violence and Children in Uganda*, Norwegian University Press, Oslo, 1987.
4. Garbarino, J., Kostelny, K., Dubrow, N., *No place to be a child: Growing up in a war zone*, Lexington Books, 1991.
5. Nixon, A., *The status of Palestinian children during the uprising in the occupied territories*. Rädda Barnen, Stockholm, 1990.
6. Netland, M., "Children and political violence." *A research report from the West Bank*, a thesis, Bergen University, Faculty of Psychology, 1991.
7. Wilson, F., Ramphele, M., "Children in South Africa." *In Children on the front line: The Impact of Apartheid, Destabilization and Warfare in Children in Southern Africa*. UNICEF, New York, 1987.
8. Macksoud, M., *How to help children affected by war: A manual for parents and teachers*. Columbia University Press, New York, 1991.
9. Boothby, N., *Helping Traumatized Children: Training Manual for Treatment and Family Reunification Program*. Save the Children. Maputo, Mozambique, 1989.
10. Raundalen, M., Dyregrov A., Bugge Gronvold R., *Reaching Children Through the Teachers. A Manual*. Norwegian and Mozambique Red Cross, Oslo and Maputo, 1990.
11. Samarasinghe, D., Nikapota, A., *Training Manual for Helping Children in Situations of Armed Conflict*, UNICEF Colombo, (draft) 1990.
12. Habash, A., *A Simple Manual in the Care of the Child*. Early Childhood Resource Centre, East Jerusalem, 1988.

CHAPTER 8

RESEARCH CHALLENGES IN PRACTICAL PERSPECTIVE

Magne Raundalen and Cole P. Dodge

Introduction

Mutual interest in children in conflict areas led to collaboration between the child psychologist and the field worker social anthropologist. The dialogue began in Uganda in the early 1980s and continued into the early '90s. We questioned whether Ugandan children were being socialized to revenge and other acts of violence and set our to investigate using both an anthropological and psychological framework. Research necessitated drawing on established methodologies and developing some new approaches as we expanded on our earlier Uganda research in the Sudan and Mozambique.

Composition writing

One of our early methods was to ask school children to write compositions. Although this has its limitations, especially in poor countries where relatively few children are literate, essays mirror the thoughts and reactions of middle and upper class children who attend school in the capital city. This method also gave us the opportunity to study the children's own views. Essay writing is a worthwhile technique, provided the teachers give as little briefing as possible and ask the children to write essays as a normal class routine. We first tried composition writing in Uganda with 650 school children in 1984-86 and reported our findings in *War, Violence and Children in Uganda*, Norwegian University Press 1987.[1]
Initially, we analyzed 450 essays on the topics of: "*War and violence in my life*", "*The story of my life*", and "*Events that made me happy and events that made me sad*". These topics evoked well written essays in which most children concentrated on personal experience of the war, describing violence ranging from loss of siblings, parents and close relatives to friends and neighbours. These essays revealed the children's understanding of their society, family and the social network in which they interacted, their fear of the military, and, finally, a good

sense of their future aspirations. Later, we asked the children in Uganda, the Sudan and Mozambique to elaborate further on the topics: "*My nation*" and "*Myself in the year 2000.*"

An early failure

An unsuccessful attempt was made in late 1986 to study Ethiopian refugee children in eastern Sudan where several NGOs were providing social services to a group of long-term refugee settlements. We worked through a medical doctor who was resident in the camp, and, he, in turn, worked through his Ethiopian counterpart responsible for education. About 300 essays were collected and sent to us from several schools in different camps with letters from the teachers providing information on the setting, tribal identification of the children and duration of stay in the camp.

We congratulated ourselves on how easy it had been to collect the essays and sat down to read and score them for various variables. By the time we had completed reading the first few essays, we confirmed our worst fears. The essays were almost identical. Every child wanted to live in either America or England and wanted to be a doctor or an engineer.

Investigation revealed that the school coordinator had briefed the teachers on what was expected and they in turn coached the children. The essays were inadequate for any type of understanding of what the refugee children thought or hoped for.

However, we did go on to use essays in southern Sudan and Mozambique. Essays from school children in Juba showed a remarkable similarity to what the children wrote in Uganda while those from Mozambique revealed a high degree of politicization among even relatively young students.

Cartoon sketches and storytelling

The problems of the war-displaced were very obvious by 1986 - even in Khartoum the capital of the Sudan and surroundings. The most obvious were the street children who roamed the city and who had been so frequently interviewed that they began to demand money and gave sarcastic answers or made up stories to those who administered the questionnaires. Street children were a recent phenomena to Khartoum and attracted the attention of many students working towards master's degrees from Khartoum University's Sociology Department. NGOs and aid agencies also commissioned research.

We assumed that if the boys could make up such fantastic stories for serious interviews, why not change the strategy and invite them to use their imagination? So, we developed a set of simple cartoon sketches and invited the children to respond to them, a technique that proved to be more successful than the ques-

Panel 8

Question: *"Do you think that street boys lose their capacity for empathy?"*

Nelson: *"I do not understand. What do you mean by that?"*

Question: *"It is often said that street boys steal, fight, become ruthless and do not care about others."*

Nelson: *"Of course we care for others. Many street children I know are responsible for the small children living alone on the street."*

Santos: *"Sometimes there are mentally ill mothers who live for a while in the street with their children. They cannot look after them properly so we have to take care of them."*

Nelson: *"Some of those children come to us for food and clothes and we have to care for them of course. We have to stop them from walking on certain streets because of the dangerous traffic..."*

Question: *"So you are responsible and caring persons?"*

Santos: *"We were taught at home, and many of us miss our smaller brothers and sisters. Some of the boys cannot talk about this because they are afraid of starting to cry."*

Question: *"But when we interview small street children they tell us that you take their money. If they get new clothes from kind people, you threaten them and strip them there and then. This is not what I call care."*

Nelson: *"Wait, wait. Small children always get some money and clothes, some. But usually not so much as we do because we are smarter. But in hard times... maybe when the newspaper has written that we are bandits with lots of money, we starve and freeze, but the small ones still get. More than before. Earlier we gave them, but they will not share now. Then we take from them if we have not been eating all day."*

Santos: *"Some of them are protected by us, but when they see how much they get, they refuse to share. That is not right."*

Interview with Nelson, age 16, and Santos, age 15, veterans of the Maputo street.

tionnaires. The cartoons were especially useful when dealing with a group who were outside the structure of a classroom. Most of the older street children were literate but they lacked the discipline to sit down and write an essay.

The cartoons were composed of a series of scenes depicting typical events that could be identified in the life of street children using a boy as the central figure. Storytelling allowed them to describe accounts of begging for money or police harassment without directly exposing themselves. The instructions with the drawings were phrased to sound like the usual text accompanying tests used in basic psychological studies from other countries. The purpose was to illicit a personal story. *"Here you see Carlos, he is alone, tell me what he is thinking, what he will do and what will happen to him."(Mozambique)* Or with the sleeping-in-the-street drawing: *"Here you see Yameen (Sudan) in the street with friends, he is sleeping, and when he sleeps he is dreaming, please tell me about his dreams and what Yameen is dreaming about?"*

The cartoon sketches had to be field tested so that the street boys could identify their own cultural setting. In our first attempt in the Sudan, the pictures were too realistic and left little to their imagination. Eventually, the sketches gave the researcher detailed and concrete information about the child's reaction to stressful episodes in terms of active-passive, fight-flight, aggression-fear and cognitive-emotional responses. The street boys provided narratives in an unrestrained way, their feelings projected onto Carlos or Yameen. They no longer had to struggle with poorly understood questions, put to them by students or social workers who were trying to fill their quota to make their research sample large enough to be manipulated for statistical significance.

Other approaches

Our approach was to gain access to the child's thinking and expectations. We created different sentence completion tests relating to the children's life situation: *"If a genie came out of the bottle and stood in front of you saying, 'I offer you three wishes,' what would your first, second and third wishes be?"* Another: *"If you were a father and had a son who ran away from home and settled in the street, what would you do then?"*[2]

The expert group

Another approach developed for the study of war-displaced and street children in Maputo incorporated the "expert group" process. When interviewing larger samples, a group of boys deemed to be reliable and from representative backgrounds, was selected. They were invited to daily follow-up meetings and engaged in a discussion of all the aspects of the information that emerged (see chapter 2).

Interviews with parents

In both Uganda and Mozambique, we interviewed parents or guardians. The purpose was simply to double-check the data obtained from their children on specific points. For example, in Mozambique, we interviewed the parent or guardian at home to verify the high proportion of reports of children who had a close relative killed. In general, this process suggested fewer such incidents but nonetheless confirmed a high level of experiences of violence and displacement. Stress was also staggeringly obvious.

Interviews and ratings by teachers and parents

Teachers and parents provided a wealth of insight. Our first approach in Uganda with a Behaviour Problem Checklist for teachers proved unproductive, compared with countries with class sizes in the 20-pupil range. In Mozambique, the Sudan and Uganda, class sizes were more typically 40 to 90 children. In Uganda and southern Sudan, we found that violence and insecurity had been so prevalent in the children's lives that it became difficult for the teacher to say which students were showing outward signs of stress. Finally, the teachers may have had less training and were, therefore, less familiar with psychological indicators. Absenteeism is chronic in many African schools for a variety of reasons and is made worse by insecurity.

Teachers in both Juba, a garrison town in southern Sudan, and Kampala, Uganda, were reluctant to show too much interest in their students because their class rooms contained students from different tribal backgrounds whose parents came from a variety of backgrounds -- some from the military and others, supporters of the guerrilla movement. A teacher who expressed too much interest in a particular student might have evoked a misunderstanding that resulted in harm. Severe reprisals were not uncommon for any one who was even suspected to be sympathetic to one side or the other. Teachers, therefore, tended to concentrate on teaching and avoided interaction which might have yielded insight to the psychological health and well being of the students.

Stress questionnaires

Stress questionnaires were used in Juba and in Mozambique after our use of them in Uganda. These were useful although somewhat tricky to administer. Translation into local languages invariably resulted in some confusion about definitive meanings. Also, when we cross-checked the children's responses with parents and teachers, we often received contradictory answers. In Mozambique, however, where the whole series of questionnaires were developed in collaboration with a local teacher and the very process of selecting and interviewing was a joint effort, much more reliable results were obtained.

Examination by clinical team

One medical doctor and a clinical psychologist examined 79 severely affected and abandoned children in a Red Cross shelter in Kampala in 1985. Even though this clinical team had an extensive check list, we were unable to use the information perhaps because the examinations they carried out were too rapid. All the records looked similar. The medical situation required rapid assessments; given the shortage of doctors and the abundance of patients.

Because of the clinical team's inability to devote sufficient time to determine the degree of stress, extensive interviews were conducted by a social worker who recorded the data. Later, the tapes were translated into English and transcribed on a word processor. These interviews provided a wealth of very good material which was used to evaluate each child and the group.

Anthropological observation

Field workers, especially those engaged in relief and emergency care, should sharpen their own awareness of the psychological aspects of stress involving violence to children as there is so little data available from Africa despite the unprecedented numbers of children who are exposed to war. Sensitive observation is potentially useful in broadening our understanding and knowledge. Relief workers are frequently the only ones who have access to children in the most difficult circumstances. For example, at the height of the civil war in Uganda, Josephine Harmsworth, a UNICEF consultant, was caught behind the lines. There she met and interviewed two child soldiers, one illiterate, the other who spoke good English, concerning their attitude about the future. The illiterate boy wanted to remain in the army while the educated child soldier planned to finish his schooling. The interviews she conducted were used by UNICEF to bolster its contention that all child soldiers should be returned to school. Later, when Yoweri Museveni came to power, these same interviews were cited in an attempt to persuade the new Head of State to have the child soldiers released and returned to school.

Finally, through anthropological participant observation, it is possible to understand cultural limits to aggression and who provides care in situations of abnormal stress. Once these are spotted, it may be relatively easy to promote therapeutic activities through just one or two key local people or institutions. What follow are four illustrative examples drawn from Uganda, the Sudan and Mozambique.

Luwero, Uganda, April 1986

We attended the official opening of the first newly installed handpump just three months after the end of the violent civil war in the Luwero Triangle. Half a million people had been killed in the preceding five years. One out of five of the local population had been displaced and all schools, health centres and hospitals closed. The NRA operated from Luwero which resulted in UNLA troops moving through entire villages shooting everything in sight -- men, women and children included.

The Prime Minister was the chief guest and a spontaneous celebration followed his speech since all handpumps had been destroyed in the war. Some schools and churches had been reopened so teachers, clergy and of course all the children for miles around were gathered to see the event. Several hundred children performed a musical play with dance, which in the course of an hour related the history of the civil war, including a mixture of well-known local incidents, and featured solo performances by children who recited their own unique agony.

Although we never found out who was responsible for this therapeutic process "staged" only three months after the overthrow of the Obote government, we were able to trace the basic idea to a weekly local radio programme of songs, poetry, plays and stories, produced by school children in Kampala from the child-to-child project. This example had given teachers and clergy the idea in Luwero. What a performance.

Maputo, Mozambique, December 1988

During the mid-1980s, between 8,000 and 10,000 youngsters were kidnapped and coerced into becoming child soldiers by the Renamo. A few hundred of these escaped or were captured by the Frelimo.

In December 1988, we visited a rehabilitation centre where 37 former child soldiers lived. The main objectives were to give them an education and to rehabilitate them for a normal life. To accomplish this, the boys were given a briefing every day about the state of the country, an analysis of the activities and atrocities of the anti-government forces. The boys were also encouraged to re-enact the atrocities they themselves had witnessed while serving with the anti-government forces. This helped to create a more positive future outlook for them. The Frelimo had a long and successful liberation struggle which involved winning popular support. The process of rehabilitating the child soldiers clearly drew upon this experience.

Khartoum, Sudan, November 1987

Thousands of boys struggle to survive on the streets of Khartoum. Most have fled the war in the south. At a dingy cross roads in the centre of town late one evening, we encountered a clutch of street boys who were settling down for the night. A small fire of refuse sputtered to life from time to time as a 12-year-old southern boy told a story. In broken Arabic, he related how a boy had fled a violent war, about his daring escape from soldiers and the cunning way he sneaked through the front lines of battle. Arriving in the capital the boy worked as an errand boy and gradually through the years became a wealthy store owner.

What, we wondered, was going on as the interpreter signalled for us to keep quiet. We later learned that this group of boys had contributed to the admission price so that one of their number could go to the cinema; they did not have enough money for everyone to go. After the lucky boy returned, he had to retell the entire story. He told it well, with action, part of the reason why he was selected in the first place. Was it a coincidence that the story had a distinct resemblance to Sudan and the war in the south? Was it possible that the movie had been made in Khartoum? No, the boy was re-casting the movie to fit the boys' setting so that the group could relate to it. Indeed the whole episode was related in a way that reflected some of the hopes and dreams of these street children. From this excursion into the world of the street children, we realized our most productive research method: cartoons and storytelling.

Nakaseki, Uganda, March 1986

A few days after the fall of Kampala to the NRA, a young girl of 15 stood at the altar of the Catholic church. She told her story to the small congregation at a special healing service to mark the reopening of the church. In a quiet voice, she told of how she and her classmates were taken from their school hostel by UNLA soldiers, then forced to walk to the barracks in Bombo where they were raped. The girls were held for several weeks until the Red Cross was able to trace their whereabouts.

She was now committing herself to God and the service of the church by entering an all girls school to prepare to become a nun. Preparing for a new place in life with the cleansing and blessing through this solemn religious ceremony. Before this, she had not been able to talk about her abduction with her parents who felt that silence was best.

Discussion

These examples illustrate a variety of culturally acceptable healing approaches using song, dance, story telling, religious ritual and political orientation. They provide insight into the ways children, who have experienced war and stress, can be helped. Undoubtedly, there are many more good examples which should be documented and encouraged.

The anxiety and Post Traumatic Stress Disorders which accompany war and insecurity have serious effects on children and society, both short-term and long-term. By using positive examples and providing a positive future for children, our purpose is to encourage more help for children in war based on observation and research, as well as by sharing new insights.[2,3]

These exploratory approaches have been used on a pilot basis to determine how research might be conducted in order to bring new insight as to how the impact of war influences children. The findings reported in this book and elsewhere were based on one or more of these approaches.

References

1. Dodge, C.P., Raundalen, M.(ed), *War, Violence and Children in Uganda,* Norwegian University Press, Oslo, 1987.

2. Innstrand, A. G., Haaseth, : "Street Children in Khartoum," Thesis University of Bergen, Norway, 1991.

3. Raundalen, M., Raundalen, T. S., Pereira, E., Vogt. N.: "Street Children in Maputo," Report to NORAD, Maputo, 1991.

CHAPTER 9

RIGHTS AND HOPES FOR CHILDREN IN WAR

Magne Raundalen and Cole P. Dodge

Article 39 and the child in war

Only recently have we admitted that war inflicts deep traumas on a child. The Convention on the Rights of the Child, adopted by the United Nations General Assembly in November 1989, and ratified in 1990, established a child's right to psychological recovery after war. The ratification of the Convention marks a major step forward in public recognition of child crisis psychology. A child's need for psychological care following trauma is recognized in article 39 of the convention: *"State parties shall take all appropriate measures to promote physical and psychological recovery and social reintegration of a child victim of: any form of neglect, exploitation, or abuse; torture or any form of cruel, inhuman or degrading treatment or punishment; or armed conflict. Such recovery and reintegration shall take place in an environment which fosters the health, self respect and dignity of the child."*

Significantly, the Convention stresses both physical and psychological recovery as well as social reintegration. But what does the phrase "all appropriate measures" mean? And how should children's rights to these be interpreted?

As far as many developed countries are concerned, the answers are clear. But many war-torn nations have also come to realize that children must receive extensive treatment for both physical and psychological wounds, particularly after war and conflict. The nations of southern Africa, Angola, Mozambique and Namibia in particular, are a case in point. The Fifth Congress of Frelimo, held in Mozambique in July 1989, considered a separate agenda item on the need to educate psychologists and teachers in the rehabilitation of war-traumatized children. There is greater awareness now that violence against - and around - children will potentially affect them and their children, for decades.[1,2]

The magnitude of the problem

However, most nations devastated by war simply do not have the resources to provide such treatment for children. It is, therefore, incumbent upon the international community to help secure the rights embodied in article 39.[3,4] The magnitude of the problem may be illustrated by a study of school children in Mozambique where we found that 75 percent had been in a situation, at least once, where they felt sure that they were going to die. In another study from the Sudan, we found that about 40 per cent of the children in our Juba sample had lost one parent.

Underestimating the sufferings

Children who experience the loss of a parent or loved one often sink deep into reactive depression. Some try to seek imaginary revenge through aggression and violence, often against those not responsible for their fate. Others may join in the war by becoming a child soldier.

Adults generally underestimate the damage done to children by the trauma of war. Journalistic accounts of children playing while conditions of war prevail, and of children's ability to spring back from violence they have experienced or witnessed, have made the work of implementing the Convention on the Rights of the Child much harder.

Avoid separation

The first preventive and therapeutic principle is that everything possible must be done to avoid separating a child from his or her family. Where separation occurs, child and family should be reunited as soon as possible.

Also essential is ensuring some semblance of normalcy in children's lives during times of crisis. If children have to be evacuated or displaced, they should be returned home as soon as possible. Naturally, physical needs must be met immediately, but where the children have been displaced for more than a few days, school routine should be re-established as a priority even though relief workers, more accustomed to life-saving medical care, may feel that schooling rates a lower priority.

Traumas 50 years later

Most revealing are studies of Finns suffering the pain of their Second World War experiences which show that many of those still disturbed had been evacuated to Sweden as children.[5] Once the war ended, they faced the desperate process of first locating and then adjusting to living with their parents. This separation and reunification process was so disturbing that many of their children -the second generation -inherited their parent's psychological wounds.

Panel 9

"Two times I have seen the bandits kill people. The first time the bandits attacked our village, my aunt was sleeping inside the house and they pulled her out. They wanted to cut her in pieces but the other one who wanted to make things easier just shot her.

"In our village, they killed seven people in the night and three were kidnapped. After this we slept in the bush, we slept there everyday for four months. In the mornings we used to go home, cook, and in the afternoons go back to the bush. The second time we were sleeping in the bush and they kidnapped people from another village. One of those kidnapped people showed where we were hiding and then they kidnapped my father because my father was from the military. I was with my mother in hiding in the bush and we saw the bandits coming with a group of people from the other village and arrived near the place we were hiding.

"One old lady had tuberculosis and she was hiding close to us. The bandits heard her when she was coughing and they discovered that there was someone there. They caught the old lady and they asked her to go with them. The lady asked to be left and the bandits killed her with knives. They divided her into two parts. When I saw this I became afraid and I started crying and then the bandits saw us and we started running away.

"That day the bandits burned houses and kidnapped people from the village. They stole everything and asked people to carry the bags with which they stole and asked them to go with the bandits. Then along the way, they were killing people. Other people managed to run away and my father was in the group that ran.

"Today I have problems with sleeping. Usually I remember when I think in the evenings and when I sleep I used to dream and when I dream I become afraid and I wake up and I start crying. And sometimes it is like I see things happening again, see the people being killed."

Eduarda, age 11, Mozambique.

Teaching the teachers

As most war-ravaged countries cannot provide children with professional therapy, teachers, health workers and religious clergy should be trained to fill the gap. Pioneering child development and child trauma psychology for child-care workers and teachers has been introduced in Mozambique. Teachers, health personal and clergy who are in close daily contact with children can be quickly taught to recognize PTSD and to address the issues of child trauma.

Talk about it

Children must be encouraged to express their pain -- verbally and through other personal means of communication. Adults often rationalize that it is best for children not to talk about traumatic experiences because they find listening too painful, but recovery can only come through communication and expression. Play, organized games, songs, art or religious rituals all potentially serve this purpose well. While frequently spontaneously generated by communities themselves, the process should be facilitated and encouraged.

Teachers can play a crucial role in helping a child to write, draw and tell his or her own story. The expression of past pain which is at the core of basic trauma treatment can come both individually and through group activity. Children must be given the opportunity for this therapy.

Paragraph 39 of the Convention on the Rights of the Child may sound abstract or too complicated to implement in a world torn apart by war. But simple and easily taught methods may ease children's pain and protect them from a self-destructive future. These methods must be brought to children in countries racked by war if we are to help break the cycle of fear and violence.

Blaming the other

In armed conflicts, the practice of blaming the other side for the sufferings of the children can be a mental exercise in relinquishing responsibility. This applies equally to the leadership or government and anti-government forces. A serious aspect, since it is unfortunately true that armed conflicts increasingly hit the civil populations including children.

It is, therefore, of the utmost importance that organizations such as UNICEF have access to leaders on both sides of the front-line as advocates for children in their efforts to commit the leaders to a new world ethic. It is also important that organizations such as the ICRC, Amnesty International and others, monitor and publish the conditions of children and point out the violation of children's rights that are taking place.

It is reassuring to note, however, that teachers are interested in seeking knowledge concerning child psychology, children's reactions and therapeutic methods. Work along these lines should continue by improving teachers' education, and using the mass media.

The distortion of the aggression issue

The issue of aggression, or the violence-breeds-violence paradigm, synonymous with war, is regularly dealt with at times of conflict, both by the press as well as the professional world, though usually in one-dimensional, poorly analyzed and superficial terms. Yet the importance of examining the children's involvement in times of war, their differing levels of participation, cannot be overstressed. For example, there is a marked difference between being the sons or daughters of peaceful peasants who become the hapless victims of guerrilla fighters compared to being heavily indoctrinated children whose hatred of the enemy can justify killing based upon religious or ethnic rivalry. The religious issue should not be downplayed, as it is all too clear that where indoctrination equates being on the side that is "good", i.e., on God's side, then the other side is defined as representative of the evil forces.

War is the ultimate expression of hatred and we should be aware of the long-term effects on the society -- the impact on culture leading to more violence. However, while it is important to tackle the human aggression issue, our research experience gives some cause for optimism: given a meaningful, constructive alternative, the great majority of the new generation will grasp it. The children we interviewed did not reveal a built-in inclination to continue to fight and to create new armed conflicts. However, we realize that this is a complicated issue and should be subject to more thinking and more research.

The concept of PTSD in children

It is now known that people who have experienced different forms of extreme stress develop similar problems. This is also evident in children and has led to the identification of a cluster of symptoms triggered by events that represent serious threats to one's own life or physical well-being; to one's close relatives and friends; and/or to the security of one's home or community.

The reaction pattern is characterized by, among others, repeatedly recollecting the traumatic event; recurrent, distressing dreams of the event; and reliving the experience with illusions, hallucinations and flashbacks.

The next group of reactions are related to feelings of intense psychological distress at exposure to events that resemble or bring to mind any aspect of the trauma, including anniversaries of the event. This often leads to persistent avoidance of stimuli associated with the trauma.

The last cluster of reactions are symptoms characterized by increased arousal and were not present before the trauma. Sleep may be impaired or the person may become irritable, have difficulty concentrating, or always be on the look-out for danger.

It is possible to list a comprehensive body of workable interventions to prevent long-term negative after-effects caused by PTSD in children. These recommendations can be followed up by teachers and augment and reinforce the methods outlined in this book; most specifically talking and working through the traumatic events as soon as possible.

Preparedness for the future

Every human being is born with the capacity to plan and think ahead. This ability is revealed very early in childhood. The child gradually becomes aware that there are situations and events far beyond the parental sphere of influence, some, such as storms, floods, and earthquakes, can be even life-threatening. The child may also discover that when it comes to war, adults may be more inclined to prepare for action rather than the prevention of war.

Recent studies[6] show that children, even at the late pre-school stage, take account of macro-events in calculating their prospects for the future. One Ugandan child wrote *"If I am not a bunch of bones in the bottom of a grave, I will plan to have an education, a job and a family."* Studies also show that traumatic events may affect a child's view of the future, creating a long-lasting negative expectation. It is, therefore, opportune and appropriate to find workable concepts which can guide a teacher's interventions to strengthen the child's capacity to believe in a constructive future.

The child politician

In studying the children's descriptions of their prospects for the future in times of war, we have come to the conclusion that the more oppressive the situation, the more political the children become. Many children invest their entire hope for the future in whether their nation can cope. Their concerns become political instead of personal. Many of them mature too early. We also observe children taking on the role of an adult or parent especially in situations where their parents show resignation, as typified by the activist children of the West Bank and black South Africa.

The positive feature of this politicalization is that it helps make the world rational, understandable and relatively predictable in their minds. The children are given an opportunity to involve themselves intellectually and even in some cases, practically, in bringing the war to an end in a constructive way. This is expressed through political demonstrations, discussions at school, at home, and in peer groups.

Where a nation's young people are politically enlightened, they are able to construct more realistic prospects for their future. Expectations will help strengthen the hope that may give them sustainable preparation for the future, through what we have termed "cognitive political intervention."

As to politics in school and political education in general, the way the Mozambican teachers have informed their pupils about the situation in southern Africa has helped counteract irrational, hostile pictures of the enemy, a contribution of the utmost importance to the peace-making process.

The concept of psycho-biography

To maintain continuity in life, which is held as a very important factor in promoting mental health, children need to communicate significant events in their life history. For this purpose we have used the concept psycho-biography both as a practical tool and as a description of an important aspect of helping war traumatized children. The features dealt with as benchmarks of the psycho-biography are transitions, stress, trauma, separations and losses. As a practical tool the field worker compiles all these issues with reference to age level when it occurred both to plan appropriate intervention and to interact with the child about crucial experiences. Thus the psycho-biography can be used as a road to mental health protocol for caregivers and teachers in much the same way as the "growth chart" is used by parents and health care workers to monitor physical growth.

The push and pull concepts

These concepts are not specifically related to children in war situations, but are drawn from our studies of street children, a phenomenon often related to war. The push-concept covers all the negative features which cause child vagrancy and ranges from war, drought, poverty, to single-parent families and others. The pull-concept is comprised of factors perceived as positive by the child which attracts him to the big city.

The importance of analysis: Since many programmes aimed at helping street children are based on reuniting them with their families, it is important to have a thorough analysis of the push-pull factors. We have met street children from war-stricken areas who have expressed mortal dread when confronted with the prospect of reunification.

Areas of hope for reaching children in war

Areas for hope in combating the overwhelming and devastating impact of civil war on ordinary civilians must begin with the newly emerging international ethic which demands that governments and UN agencies respond to the loud emergencies brought about by conflict. Tied to the emergence of an ethic of response is the role of the media which bring reports and pictures from the remotest corners of Africa into the homes of millions in almost all countries of the world. Media coverage contributes to the new ethic by making the world a smaller place and uniting people in their collective humanitarian outrage against atrocities and suffering to demand action by their respective governments. The power of the media is readily seen from the acute attentiveness which government and liberation movements alike give to international news about conflict, insecurity and civil war.

Reaching children across the front lines

A further area of hope has been the success of Operation Lifeline Sudan[7,8] which in 1989 delivered food, medical supplies and other relief items to southern Sudan for the first time to both sides of the conflict. Also, the corridor of peace in Uganda in late 1985 and early 1986 which operated with the consent of both the Kampala authorities and the National Resistance Army headquartered in Kasese and delivered medical supplies from Kampala directly to areas not under government control. Both these examples represent the ability to provide relief to victims of war and provide an important precedent of consent from both sides. Similarly, food deliveries from Addis Ababa to Tigre Province in northern Ethiopia reflects the same international and media insistence that relief be given to civilians.

Making a reality of the Convention

The Convention on the Rights of the Child, while focusing specifically on children, provides governments, UN agencies, regional associations such as the Organization for African Unity, the ICRC, human rights organizations and NGOs as well as the media, with a persuasive new international instrument to reach children caught in war situations. Children account for nearly half the population of African countries and are the most vulnerable along with their mothers and grandparents during wartime.

A less tense world situation

The radical political changes sweeping Eastern Europe have global implications for limiting big power support to civil war within Africa. With less support from abroad, African governments may be more likely to negotiate internal disputes.

However, declining geopolitical importance may cost the countries of Africa social and humanitarian aid, imperiling further their already precarious populations. Humanitarian and social development aid levels have followed geopolitical commitments in the past. Events in Eastern Europe may well absorb additional amounts from relatively fixed foreign aid allocations. This would imply less aid for Africa. Similarly, the Gulf war has consumed massive amounts of money at a time when the world economy is weak, which could well result in high fuel import bills and reductions in aid to Africa.

The reallocation of money

One important side benefit to the lessening of international tension is disarmament. With $1.8 million spent every minute for global military purposes (not including the Gulf war), it is possible to envision that the decade of the 1990s will see less money invested in the arms race or for defence purposes and hope that some of the resultant savings will be invested in the health and well-being of people in the world's least developed countries. The worldwide budget of UNICEF accounted for four hours and twenty six minutes of global military expenditure in 1987. The Brandt Commission noted that "...*the military expenditure of only half a day would suffice to finance the whole malaria eradication programme of the World Health Organization...*" And that "*... for the price of one jet fighter ($21 million) one could set up about 40,000 village pharmacies.*"

Children want peace

A final source of hope can, we think, be derived from the interviews and meetings with hundreds of children in war-torn Uganda, the Sudan and Mozambique, the overwhelming majority of whom consistently expressed themselves to be tired of war. They have hopes and dreams for their own futures. They want to be gainfully employed, get married, have children, a secure home, the ability to look after their parents, and good health and education for their own children. They are not at all preoccupied by revenge or any other prospects that could prolong the war situation. These children are the leaders of tomorrow. They dare to hope.

References

1. Palme, L., "Why the Children's Convention?" Speech given at a UNICEF and Swedish Save the Children Conference on the Convention on the Rights of the Child, October, 1988.
2. Palme, L., "Adults Responding to Children in Times of War." Speech at The Seminar of the NGO Committee on UNICEF, April, New York, 1991.
3. Kent, George, *Implementing the Rights of Children in Armed Conflict*, (draft), University of Hawaii, Honolulu, 1991.
4. Cohn, Ilene, "The Convention on the Rights of the Child: What it means for Children in War." *International Journal of Refugee Law*, Vol. 3., No. 1, 1991.
5. Lagnebro, L. : *Finnish Children of War*, Institute of Migration, Turkul, Finland, (in press), 1991.
6. Dodge, C.P., Raundalen, M.(ed), *War, Violence and Children in Uganda*, Norwegian University Press, 1987.
7. Minear, L., *Humanitarianism Under Seige, A Critical Review of Operation Lifeline Sudan*, Red Sea Press Inc., New Jersey, 1991.

Convention on the rights of the child

PREAMBLE

The States Parties to the present Convention,

Considering that, in accordance with the principles proclaimed in the Charter of the United Nations, recognition of the inherent dignity and of the equal and inalienable rights of all members of the human family is the foundation of freedom, justice and peace in the world,

Bearing in mind that the peoples of the United Nations have, in the Charter, reaffirmed their faith in fundamental human rights and in the dignity and worth of the human person, and have determined to promote social progress and better standards of life in larger freedom,

Recognizing that the United Nations has, in the Universal Declaration of Human Rights and in the International Convenants on Human Rights, proclaimed and agreed that everyone is entitled to all the rights and freedoms set forth therein, without distinction of any kind, such as race, colour, sex, language, religion, political or other opinion, national or social origin, property, birth or other status,

Recalling that, in the Universal Declaration of Human Rights, the United Nations has proclaimed that childhood is entitled to special care and assistance,

Convinced that the family, as the fundamental group of society and the natural environment for the growth and well-being of all its members and particularly children, should be afforded the necessary protection and assistance so that it can fully assume its responsibilities within the community,

Recognizing that the child, for the full and harmonious development of his or her personality, should grow up in a family environment, in an atmosphere of happiness, love and understanding,

Considering that the child should be fully prepared to live an individual life in society, and brought up in the spirit of the ideals proclaimed in the Charter of the United Nations, and in particular in the spirit of peace, dignity, tolerance, freedom, equality and solidarity,

Bearing in mind that the need to extend particular care to the child has been stated in the Geneva Declaration of the Rights of the Child of 1924 and in the Declaration of the Rights of the Child adopted by the General Assembly on 20 November 1959 and recognized in the Universal Declaration of Human Rights, in the International Covenant on Civil and Political Rights (in particular in articles 23 and 24), in the International Covenant on Economic, Social and Cultural Rights (in particular in article 10) and in the statutes and relevant instruments of specialized agencies and international organizations concerned with the welfare of children,

Bearing in mind that, as indicated in the Declaration of the Rights of the Child, "the child, by reason of his physical and mental immaturity, needs special safeguards and care, including appropriate legal protection, before as well as after birth",

Recalling the provisions of the Declaration on Social and Legal Principles relating to the Protection and Welfare of Children, with Special Reference to Foster Placement and Adoption Nationally and Internationally; the United Nations Standard Minimum Rules for the Administration of Juvenile Justice (The Beijing Rules); and the Declaration on the Protection of Women and Children in Emergency and Armed Conflict,

Recognizing that, in all countries in the world, there are children living in exceptionally difficult conditions, and that such children need special consideration,

Taking due account of the importance of the traditions and cultural values of each people for the protection and harmonious development of the child,

Recognizing the importance of international co-operation for improving the living conditions of children in every country, in particular in the developing countries,

Have agreed as follows:

PART I
Article 1
For the purposes of the present Convention, a child means every human being below the age of eighteen years unless, under the law applicable to the child, majority is attained earlier.

Article 2
1. States Parties shall respect and ensure the rights set forth in the present Convention to each child within their jurisdiction without discrimination of any kind, irrespective of the child's or his or her parent's or legal guardian's race, colour, sex, language, religion, political or other opinion, national, ethnic or social origin, property, disability, birth or other status.

2. States Parties shall take all appropriate measures to ensure that the child is protected against all forms of discrimination or punishment on the basis of the status, activities, expressed opinions, or beliefs of the child's parents, legal guardians, or family members.

Article 3
1. In all actions concerning children, whether undertaken by public or private social welfare institutions, courts of law, administrative authorities or legislative bodies, the best interests of the child shall be a primary consideration.

2. States Parties undertake to ensure the child such protection and care as is necessary for his or her well-being, taking into account the rights and duties of his or her parents, legal guardians, or other individuals legally responsible for him or her, and, to this end, shall take all appropriate legislative and administrative measures.

3. States Parties shall ensure that the institutions, services and facilities responsible for the care or protection of children shall conform with the standards established by competent authorities, particularly in the areas of safety, health, in the number and suitability of their staff, as well as competent supervision.

Article 4
States Parties shall undertake all appropriate legislative, administrative, and other measures for the implementation of the rights recognized in the present Convention. With regard to economic, social and cultural rights, States Parties shall undertake such measures to the maximum extent of their available resources and, where needed, within the framework of international co-operation.

Article 5

States Parties shall respect the responsibilities, rights and duties of parents or, where applicable, the members of the extended family or community as provided for by local custom, legal guardians or other persons legally responsible for the child, to provide, in a manner consistent with the evolving capacities of the child, appropriate direction and guidance in the exercise by the child of the rights recognized in the present Convention.

Article 6

1. States Parties recognize that every child has the inherent right to life.

2. States Parties shall ensure to the maximum extent possible the survival and development of the child.

Article 7

1. The child shall be registered immediately after birth and shall have the right from birth to a name, the right to acquire a nationality and, as far as possible, the right to know and be cared for by his or her parents.

2. States Parties shall ensure the implementation of these rights in accordance with their national law and their obligations under the relevant international instruments in this field, in particular where the child would otherwise be stateless.

Article 8

1. States Parties undertake to respect the right of the child to preserve his or her identity, including nationality, name and family relations as recognized by law without unlawful interference.

2. Where a child is illegally deprived of some or all of the elements of his or her identity, States Parties shall provide appropriate assistance and protection, with a view to speedily re-establishing his or her identity.

Article 9

1. States Parties shall ensure that a child shall not be separated from his or her parents against their will, except when competent authorities subject to judicial review determine, in accordance with applicable law and procedures, that such separation is necessary for the best interests of the child. Such determination may be necessary in a particular case such as one involving abuse or neglect of the child by the parents, or one where the parents are living separately and a decision must be made as to the child's place of residence.

2. In any proceedings pursuant to paragraph 1 of the present article, all interested parties shall be given an opportunity to participate in the proceedings and make their views known.

3. States Parties shall respect the right of the child who is separated from one or both parents to maintain personal relations and direct contact with both parents on a regular basis, except if it is contrary to the child's best interests.

4. Where such separation results from any action initiated by a State Party, such as the detention, imprisonment, exile, deportation or death (including death arising from any cause while the person is in the custody of the State) of one or both parents or of the child, that State Party shall, upon request, provide the parents, the child or, if appropriate, another member of the family with the essential information concerning the whereabouts of the absent member(s) of the family unless the provision of the information would be detrimental to the well-being of the child. States Parties shall further ensure that the submission of such a request shall of itself entail no adverse consequences for the person(s) concerned.

Article 10

1. In accordance with the obligation of States Parties under article 9, paragraph 1, applications by a child or his or her parents to enter or leave a State Party for the purpose of family reunification shall be dealt with by States Parties in a positive, humane and expeditious manner. States Parties shall further ensure that the submission of such a request shall entail no adverse consequences for the applicants and for the members of their family.

2. A child whose parents reside in different States shall have the right to maintain on a regular basis, save in exceptional circumstances personal relations and direct contacts with both parents. Towards that end and in accordance with the obligation of States Parties under article 9, paragraph 1, States Parties shall respect the right of the child and his or her parents to leave any country, including their own, and to enter their own country. The right to leave any country shall be subject only to such restrictions as are prescribed by law and which are necessary to protect the national security, public order *(ordre public)*, public health or morals or the rights and freedoms of others and are consistent with the other rights recognized in the present Convention.

Article 11

1.　States Parties shall take measures to combat the illicit transfer and non-return of children abroad.

2.　To this end, States Parties shall promote the conclusion of bilateral or multilateral agreements or accession to existing agreements.

Article 12

1.　States Parties shall assure to the child who is capable of forming his or her own views the right to express those views freely in all matters affecting the child, the views of the child being given due weight in accordance with the age and maturity of the child.

2.　For this purpose, the child shall in particular be provided the opportunity to be heard in any judicial and administrative proceedings affecting the child, either directly, or through a representative or an appropriate body, in a manner consistent with the procedural rules of national law.

Article 13

1.　The child shall have the right to freedom of expression; this right shall include freedom to seek, receive and impart information and ideas of all kinds, regardless of frontiers, either orally, in writing or in print, in the form of art, or through any other media of the child's choice.

2.　The exercise of this right may be subject to certain restrictions, but these shall only be such as are provided by law and are necessary:

　　a)　For respect of the rights or reputations of others; or

　　b)　For the protection of national security or of public order *(ordre public)*, or of public health or morals.

Article 14

1.　States Parties shall respect the right of the child to freedom of thought, conscience and religion.

2.　States Parties shall respect the rights and duties of the parents and, when applicable, legal guardians, to provide direction to the child in the exercise of his or her right in a manner consistent with the evolving capacities of the child.

3. Freedom to manifest one's religion or beliefs may be subject only to such limitations as are prescribed by law and are necessary to protect public safety, order, health or morals, or the fundamental rights and freedoms of others.

Article 15

1. States Parties recognize the rights of the child to freedom of association and to freedom of peaceful assembly.

2. No restrictions may be placed on the exercise of these rights other than those imposed in conformity with the law and which are necessary in a democratic society in the interests of national security or public safety, public order *(ordre public)*, the protection of public health or morals or the protection of the rights and freedoms of others.

Article 16

1. No child shall be subjected to arbitrary or unlawful interference with his or her privacy, family, home or correspondence, nor to unlawful attacks on his or her honour and reputation.

2. The child has the right to the protection of the law against such interference or attacks.

Article 17

States Parties recognize the important function performed by the mass media and shall ensure that the child has access to information and material from a diversity of national and international sources, especially those aimed at the promotion of his or her social, spiritual and moral well-being and physical and mental health. To this end, States Parties shall:

   a) Encourage the mass media to disseminate information and material of social and cultural benefit to the child and in accordance with the spirit of article 29;

   b) Encourage international co-operation in the production, exchange and dissemination of such information and material from a diversity of cultural, national and international sources;

   c) Encourage the production and dissemination of children's books;

   d) Encourage the mass media to have particular regard to the linguistic needs of the child who belongs to a minority group or who is indigenous;

e) Encourage the development of appropriate guidelines for the protection of the child from information and material injurious to his or her well-being, bearing in mind the provisions of articles 13 and 18.

Article 18

1. States Parties shall use their best efforts to ensure recognition of the principle that both parents have common responsibilities for the upbringing and development of the child. Parents or, as the case may be, legal guardians, have the primary responsibility for the upbringing and development of the child. The best interests of the child will be their basic concern.

2. For the purpose of guaranteeing and promoting the rights set forth in the present Convention, States Parties shall render appropriate assistance to parents and legal guardians in the performance of their child-rearing responsibilities and shall ensure the development of institutions, facilities and services for the care of children.

3. States Parties shall take all appropriate measures to ensure that children of working parents have the right to benefit from child-care services and facilities for which they are eligible.

Article 19

1. States parties shall take all appropriate legislative, administrative, social and educational measures to protect the child from all forms of physical or mental violence, injury or abuse, neglect or negligent treatment, maltreatment or exploitation, including sexual abuse, while in the care of parent(s), legal guardian(s) or any other person who has the care of the child.

2. Such protective measures should, as appropriate, include effective procedures for the establishment of social programmes to provide necessary support for the child and for those who have the care of the child, as well as for other forms of prevention and for identification, reporting, referral, investigation, treatment and follow-up of instances of child maltreatment described heretofore, and, as appropriate, for judicial involvement.

Article 20

1. A child temporarily or permanently deprived of his or her family environment, or in whose own best interests cannot be allowed to remain in that environment, shall be entitled to special protection and assistance provided by the State.

2. States Parties shall in accordance with their national laws ensure alternative care for such a child.

3. Such care could include, *inter alia,* foster placement, *kafalah* of Islamic law, adoption or if necessary placement in suitable institutions for the care of children. When considering solutions, due regard shall be paid to the desirability of continuity in a child's upbringing and to the child's ethnic, religious, cultural and linguistic background.

Article 21

States Parties that recognize and/or permit the system of adoption shall ensure that the best interests of the child shall be the paramount consideration and they shall:

a) Ensure that the adoption of a child is authorized only by competent authorities who determine, in accordance with applicable law and procedures and on the basis of all pertinent and reliable information, that the adoption is permissible in view of the child's status concerning parents, relatives and legal guardians and that, if required, the persons concerned have given their informed consent to the adoption on the basis of such counselling as may be necessary;

b) Recognize that inter-country adoption may be considered as an alternative means of child's care, if the child cannot be placed in a foster or an adoptive family or cannot in any suitable manner be cared for in the child's country of origin;

(c) Ensure that the child concerned by inter-country adoption enjoys safeguards and standards equivalent to those existing in the case of national adoption;

d) Take all appropriate measures to ensure that, in inter-country adoption, the placement does not result in improper financial gain for those involved in it;

e) Promote, where appropriate, the objectives of the present article by concluding bilateral or multilateral arrangements or agreements, and endeavour, within this framework, to ensure that the placement of the child in another country is carried out by competent authorities or organs.

Article 22

1. States Parties shall take appropriate measures to ensure that a child who is seeking refugee status or who is considered a refugee in accordance with applicable international or domestic law and procedures shall, whether unaccompanied or accompanied by his or her parents or by any other person, receive appropriate protection and humanitarian assistance in the enjoyment of applicable rights set forth in the present Convention and in other international human rights or humanitarian instruments to which the said States are Parties.

2. For this purpose, States Parties shall provide, as they consider appropriate, co-operation in any efforts by the United Nations and other competent intergovernmental organizations or non-governmental organizations co-operating with the United Nations to protect and assist such a child and to trace the parents or other members of the family of any refugee child in order to obtain information necessary for reunification with his or her family. In cases where no parents or other members of the family can be found, the child shall be accorded the same protection as any other child permanently or temporarily deprived of his or her family environment for any reason, as set forth in the present Convention.

Article 23

1. States Parties recognize that a mentally or physically disabled child should enjoy a full and decent life, in conditions which ensure dignity, promote self-reliance and facilitate the child's active participation in the community.

2. States Parties recognize the right of the disabled child to special care and shall encourage and ensure the extension, subject to available resources, to the eligible child and those responsible for his or her care, of assistance for which application is made and which is appropriate to the child's condition and to the circumstances of the parents or others caring for the child.

3. Recognizing the special needs of a disabled child, assistance extended in accordance with paragraph 2 of the present article shall be provided free of charge, whenever possible, taking into account the financial resources of the parents or others caring for the child, and shall be designed to ensure that the disabled child has effective access to and receives education, training, health care services, rehabilitation services, preparation for employment and recreation opportunities in a manner conducive to the child's achieving the fullest possible social integration and individual development, including his or her cultural and spiritual development.

4. States Parties shall promote, in the spirit of international co-operation, the exchange of appropriate information in the field of preventive health care and of medical, psychological and functional treatment of disabled children, including dissemination of and access to information concerning methods of rehabilitation, education and vocational services, with the aim of enabling States Parties to improve their capabilities and skills and to widen their experience in these areas. In this regard, particular account shall be taken of the needs of developing countries.

Article 24

1. States Parties recognize the right of the child to the enjoyment of the highest attainable standard of health and to facilities for the treatment of illness and rehabilitation of health. States Parties shall strive to ensure that no child is deprived of his or her right of access to such health care services.

2. States Parties shall pursue full implementation of this right and, in particular, shall take appropriate measures:

a) To diminish infant and child mortality;

b) To ensure the provision of necessary medical assistance and health care to all children with emphasis on the development of primary health care;

c) To combat disease and malnutrition, including within the framework of primary health care, through, *inter alia,* the application of readily available technology and through the provision of adequate nutritious foods and clean drinking-water, taking into consideration the dangers and risks of environmental pollution;

d) To ensure appropriate pre-natal and post-natal health care for mothers;

e) To ensure that all segments of society, in particular parents and children, are informed, have access to education and are supported in the use of basic knowledge of child health and nutrition, the advantages of breast-feeding, hygiene and environmental sanitation and the prevention of accidents;

f) To develop preventive health care, guidance for parents and family planning education and services.

3. States Parties shall take all effective and appropriate measures with a view to abolishing traditional practices prejudicial to the health of children.

4. States Parties undertake to promote and encourage international co-operation with a view to achieving progressively the full realization of the right recognized in the present article. In this regard, particular account shall be taken of the needs of developing countries.

Article 25

States Parties recognize the right of a child who has been placed by the competent authorities for the purposes of care, protection or treatment of his or her physical or mental health, to a periodic review of the treatment provided to the child and all other circumstances relevant to his or her placement.

Article 26

1. States Parties shall recognize for every child the right to benefit from social security, including social insurance, and shall take the necessary measures to achieve the full realization of this right in accordance with their national law.

2. The benefits should, where appropriate, be granted, taking into account the resources and the circumstances of the child and persons having responsibility for the maintenance of the child, as well as any other consideration relevant to an application for benefits made by or on behalf of the child.

Article 27

1. States Parties recognize the right of every child to a standard of living adequate for the child's physical, mental, spiritual, moral and social development.

2. The parent(s) or others responsible for the child have the primary responsibility to secure, within their abilities and financial capacities, the conditions of living necessary for the child's development.

3. States Parties, in accordance with national conditions and within their means, shall take appropriate measures to assist parents and others responsible for the child to implement this right and shall in case of need provide material assistance and support programmes, particularly with regard to nutrition, clothing and housing.

4. States Parties shall take all appropriate measures to secure the recovery of maintenance for the child from the parents or other persons having financial responsibility for the child, both within the State Party and from abroad. In particular, where the person having financial responsibility for the child lives in a State different from that of the child, States Parties shall promote the accession to international agreements or the conclusion of such agreements, as well as the making of other appropriate arrangements.

Article 28

1. States Parties recognize the right of the child to education, and with a view to achieving this right progressively and on the basis of equal opportunity, they shall, in particular:

   a) Make primary education compulsory and available free to all;

   b) Encourage the development of different forms of secondary education, including general and vocational education, make them available and accessible to every child, and take appropriate measures such as the introduction of free education and offering financial assistance in case of need;

   c) Make higher education accessible to all on the basis of capacity by every appropriate means;

   d) Make educational and vocational information and guidance available and accessible to all children;

   e) Take measures to encourage regular attendance at schools and the reduction of drop-out rates.

2. States Parties shall take all appropriate measures to ensure that school discipline is administered in a manner consistent with the child's human dignity and in conformity with the present Convention.

3. States Parties shall promote and encourage international co-operation in matters relating to education, in particular with a view to contributing to the elimination of ignorance and illiteracy throughout the world and facilitating access to scientific and technical knowledge and modern teaching methods. In this regard, particular account shall be taken of the needs of developing countries.

Article 29

1. States Parties agree that the education of the child shall be directed to:

   a) The development of the child's personality, talents and mental and physical abilities to their fullest potential;

   b) The development of respect for human rights and fundamental freedoms, and for the principles enshrined in the Charter of the United Nations;

   c) The development of respect for the child's parents, his or her own cultural identity, language and values, for the national values of the country in which the child is living, the country from which he or she may originate, and for civilizations different from his or her own;

   d) The preparation of the child for responsible life in a free society, in the spirit of understanding, peace, tolerance, equality of sexes, and friendship among all peoples, ethnic, national and religious groups and persons of indigenous origin;

   e) The development of respect for the natural environment.

2. No part of the present article or article 28 shall be construed so as to interfere with the liberty of individuals and bodies to establish and direct educational institutions, subject always to the observance of the principles set forth in paragraph 1 of the present article and to the requirements that the education given in such institutions shall conform to such minimum standards as may be laid down by the State.

Article 30

In those States in which ethnic, religious or linguistic minorities or persons of indigenous origin exist, a child belonging to such a minority or who is indigenous shall not be denied the right, in community with other members of his or her group, to enjoy his or her own culture, to profess and practise his or her own religion, or to use his or her own language.

Article 31

1. States Parties recognize the right of the child to rest and leisure, to engage in play and recreational activities appropriate to the age of the child and to participate freely in cultural life and the arts.

2. States Parties shall respect and promote the right of the child to participate fully in cultural and artistic life and shall encourage the provision of appropriate and equal opportunities for cultural, artistic, recreational and leisure activity.

Article 32

1. States Parties recognize the right of the child to be protected from economic exploitation and from performing any work that is likely to be hazardous or to interfere with the child's education, or to be harmful to the child's health or physical, mental, spiritual, moral or social development.

2. States parties shall take legislative, administrative, social and educational measures to ensure the implementation of the present article. To this end, and having regard to the relevant provisions of other international instruments, States Parties shall in particular:

> a) Provide for a minimum age or minimum ages for admission to employment;
>
> b) Provide for appropriate regulation of the hours and conditions of employment;
>
> c) Provide for appropriate penalties or other sanctions to ensure the effective enforcement of the present article.

Article 33

States Parties shall take all appropriate measures, including legislative, administrative, social and educational measures, to protect children from the illicit use of narcotic drugs and psychotropic substances as defined in the relevant international treaties, and to prevent the use of children in the illicit production and trafficking of such substances.

Article 34

States Parties undertake to protect the child from all forms of sexual exploitation and sexual abuse. For these purposes, States Parties shall in particular take all appropriate national, bilateral and multilateral measures to prevent:

> a) The inducement or coercion of a child to engage in any unlawful sexual activity;
>
> b) The exploitative use of children in prostitution or other unlawful sexual practices;
>
> c) The exploitative use of children in pornographic performances and materials.

Article 35

States Parties shall take all appropriate national, bilateral and multilateral measures to prevent the abduction of, the sale of or traffic in children for any purpose or in any form.

Article 36

States Parties shall protect the child against all other forms of exploitation prejudicial to any aspects of the child's welfare.

Article 37

States Parties shall ensure that:

    a) No child shall be subjected to torture or other cruel, inhuman or degrading treatment or punishment. Neither capital punishment nor life imprisonment without possibility of release shall be imposed for offences committed by persons below eighteen years of age;

    b) No child shall be deprived of his or her liberty unlawfully or arbitrarily. The arrest, detention or imprisonment of a child shall be in conformity with the law and shall be used only as a measure of last resort and for the shortest appropriate period of time;

    c) Every child deprived of liberty shall be treated with humanity and respect for the inherent dignity of the human person, and in a manner which takes into account the needs of persons of his or her age. In particular, every child deprived of liberty shall be separated from adults unless it is considered in the child's best interest not to do so and shall have the right to maintain contact with his or her family through correspondence and visits, save in exceptional circumstances;

    d) Every child deprived of his or her liberty shall have the right to prompt access to legal and other appropriate assistance, as well as the right to challenge the legality of the deprivation of his or her liberty before a court or other competent, independent and impartial authority, and to a prompt decision on any such action.

Article 38

1. States Parties undertake to respect and to ensure respect for rules of international humanitarian law applicable to them in armed conflicts which are relevant to the child.

2. States Parties shall take all feasible measures to ensure that persons who have not attained the age of fifteen years do not take a direct part in hostilities.

3. States Parties shall refrain from recruiting any person who has not attained the age of fifteen years into their armed forces. In recruiting among those persons who have attained the age of fifteen years but who have not attained the age of eighteen years, States Parties shall endeavour to give priority to those who are oldest.

4. In accordance with their obligations under international humanitarian law to protect the civilian population in armed conflicts, States Parties shall take all feasible measures to ensure protection and care of children who are affected by an armed conflict.

Article 39

States Parties shall take all appropriate measures to promote physical and psychological recovery and social reintegration of a child victim of: any form of neglect, exploitation, or abuse; torture or any other form of cruel, inhuman or degrading treatment or punishment; or armed conflicts. Such recovery and reintegration shall take place in an environment which fosters the health, self-respect and dignity of the child.

Article 40

1. States Parties recognize the right of every child alleged as, accused of, or recognized as having infringed the penal law to be treated in a manner consistent with the promotion of the child's sense of dignity and worth, which reinforces the child's respect for the human rights and fundamental freedoms of others and which takes into account the child's age and the desirability of promoting the child's reintegration and the child's assuming a constructive role in society.

2. To this end, and having regard to the relevant provisions of international instruments, States Parties shall, in particular, ensure that:

>   a) No child shall be alleged as, be accused of, or recognized as having infringed the penal law by reason of acts or omissions that were not prohibited by national or international law at the time they were committed;

b) Every child alleged as or accused of having infringed the penal law has at least the following guarantees:

i) To be presumed innocent until proven guilty according to law;

ii) To be informed promptly and directly of the charges against him or her, and, if appropriate, through his or her parents or legal guardians, and to have legal or other appropriate assistance in the preparation and presentation of his or her defence;

iii) To have the matter determined without delay by a competent, independent and impartial authority or judicial body in a fair hearing according to law, in the presence of legal or other appropriate assistance and, unless it is considered not to be in the best interest of the child, in particular, taking into account his or her age or situation, his or her parents or legal guardians;

iv) Not to be compelled to give testimony or to confess guilt; to examine or have examined adverse witnesses and to obtain the participation and examination of witnesses on his or her behalf under conditions of equality;

v) If considered to have infringed the penal law, to have this decision and any measures imposed in consequence thereof reviewed by a higher competent, independent and impartial authority or judicial body according to law;

vi) To have the free assistance of an interpreter if the child cannot understand or speak the language used;

vii) To have his or her privacy fully respected at all stages of the proceedings.

3. States Parties shall seek to promote the establishment of laws, procedures, authorities and institutions specifically applicable to children alleged as, accused of, or recognized as having infringed the penal law, and, in particular:

a) The establishment of a minimum age below which children shall be presumed not to have the capacity to infringe the penal law;

b) Whenever appropriate and desirable, measures for dealing with such children without resorting to judicial proceeding, providing that human rights and legal safeguards are fully respected.

4. A variety of dispositions, such as care, guidance and supervision orders; counselling; probation; foster care; education and vocational training programmes and other alternatives to institutional care shall be available to ensure that children are dealt with in a manner appropriate to their well-being and proportionate both to their circumstances and the offence.

Article 41

Nothing in the present Convention shall affect any provisions which are more conducive to the realization of the rights of the child and which may be contained in:

a) The law of a State Party; or

b) International law in force for that State.

PART II

Article 42

States Parties undertake to make the principles and provisions of the Convention widely known, by appropriate and active means, to adults and children alike.

Article 43

1. For the purpose of examining the progress made by States Parties in achieving the realization of the obligations undertaken in the present Convention, there shall be established a Committee on the Rights of the Child, which shall carry out the functions hereinafter provided.

2. The Committee shall consist of ten experts of high moral standing and recognized competence in the field covered by this Convention. The members of the Committee shall be elected by States Parties from among their nationals and shall serve in their personal capacity, consideration being given to equitable geographical distribution, as well as to the principal legal systems.

3. The members of the Committee shall be elected by secret ballot from a list of persons nominated by States Parties. Each State Party may nominate one person from among its own nationals.

4. The initial election to the Committee shall be held no later than six months after the date of the entry into force of the present Convention and thereafter every second year. At least four months before the date of each election, the Secretary-General of the United Nations shall address a letter

to States Parties inviting them to submit their nominations within two months. The Secretary-General shall subsequently prepare a list in alphabetical order of all persons thus nominated, indicating States Parties which have nominated them, and shall submit it to the States Parties to the present Convention.

5.   The elections shall be held at meetings of States Parties convened by the Secretary-General at United Nations Headquarters. At those meetings, for which two thirds of States Parties shall constitute a quorum, the persons elected to the Committee shall be those who obtain the largest number of votes and an absolute majority of the votes of the representatives of States Parties present and voting.

6.   The members of the Committee shall be elected for a term of four years. They shall be eligible for re-election if renominated. The term of five of the members elected at the first election shall expire at the end of two years; immediately after the first election, the names of these five members shall be chosen by lot by the Chairman of the meeting.

7.   If a member of the Committee dies or resigns or declares that for any other cause he or she can no longer perform the duties of the Committee, the State Party which nominated the member shall appoint another expert from among its nationals to serve the remainder of the term, subject to the approval of the Committee.

8.   The Committee shall establish its own rules of procedure.

9.   The Committee shall elect its officers for a period of two years.

10.   The meetings of the Committee shall normally be held at United Nations Headquarters or at any other convenient place as determined by the Committee. The Committee shall normally meet annually. The duration of the meetings of the Committee shall be determined, and reviewed, if necessary, by a meeting of the States Parties to the present Convention, subject to the approval of the General Assembly.

11.   The Secretary-General of the United Nations shall provide the necessary staff and facilities for the effective performance of the functions of the Committee under the present Convention.

12.   With the approval of the General Assembly, the members of the Committee established under the present Convention shall receive emoluments from United Nations resources on such terms and conditions as the Assembly may decide.

Article 44

1. States Parties undertake to submit to the Committee, through the Secretary-General of the United Nations, reports on the measures they have adopted which give effect to the rights recognized herein and on the progress made on the enjoyment of those rights:

 a) Within two years of the entry into force of the Convention for the State Party concerned;

 b) Thereafter every five years.

2. Reports made under the present article shall indicate factors and difficulties, if any, affecting the degree of fulfillment of the obligations under the present Convention. Reports shall also contain sufficient information to provide the Committee with a comprehensive understanding of the implementation of the Convention in the country concerned.

3. A State Party which has submitted a comprehensive initial report to the Committee need not, in its subsequent reports submitted in accordance with paragraph 1 (b) of the present article, repeat basic information previously provided.

4. The Committee may request from States Parties further information relevant to the implementation of the Convention.

5. The Committee shall submit to the General Assembly, through the Economic and Social Council, every two years, reports on its activities.

6. States Parties shall make their reports widely available to the public in their own countries.

Article 45

In order to foster the effective implementation of the Convention and to encourage international co-operation in the field covered by the Convention:

 a) The specialized agencies, the United Nations Children's Fund, and other United Nations organs shall be entitled to be represented at the consideration of the implementation of such provisions of the present Convention as fall within the scope of their mandate. The Committee may invite the specialized agencies, the United Nations Children's Fund and other competent bodies as it may consider appropriate to provide expert advice on the implementation of the

Convention in areas falling within the scope of their respective mandates. The Committee may invite the specialized agencies, the United Nations Children's Fund, and other United Nations organs to submit reports on the implementation of the Convention in areas falling within the scope of their activities;

b) The Committee shall transmit, as it may consider appropriate, to the specialized agencies, the United Nations Children's Fund and other competent bodies, any reports from States Parties that contain a request, or indicate a need, for technical advice or assistance, along with the Committee's observations and suggestions, if any, on these requests or indications;

c) The Committee may recommend to the General Assembly to request the Secretary-General to undertake on its behalf studies on specific issues relating to the rights of the child;

d) The Committee may make suggestions and general recommendations based on information received pursuant to articles 44 and 45 of the present Convention. Such suggestions and general recommendations shall be transmitted to any State Party concerned and reported to the General Assembly, together with comments, if any, from State Parties.

PART III

Article 46

The present Convention shall be open for signature by all States.

Article 47

The present Convention is subject to ratification. Instruments of ratification shall be deposited with the Secretary-General of the United Nations.

Article 48

The present Convention shall remain open for accession by any State. The instruments of accession shall be deposited with the Secretary-General of the United Nations.

Article 49

1. The present Convention shall enter into force on the thirtieth day following the date of deposit with the Secretary-General of the United Nations of the twentieth instrument of ratification or accession.

2. For each State ratifying or acceding to the Convention after the deposit of the twentieth instrument of ratification or accession, the Convention shall enter into force on the thirtieth day after the deposit by such State of its instrument of ratification or accession.

Article 50

1. Any State Party may propose an amendment and file it with the Secretary-General of the United Nations. The Secretary-General shall thereupon communicate the proposed amendment to States Parties, with a request that they indicate whether they favour a conference of States Parties for the purpose of considering and voting upon the proposals. In the event that, within four months from the date of such communication, at least one third of the States Parties favour such a conference, the Secretary-General shall convene the conference under the auspices of the United Nations. Any amendment adopted by a majority of States Parties present and voting at the conference shall be submitted to the General Assembly for approval.

2. An amendment adopted in accordance with paragraph 1 of the present article shall enter into force when it has been approved by the General Assembly of the United Nations and accepted by a two-thirds majority of States Parties.

3. When an amendment enters into force, it shall be binding on those States Parties which have accepted it, other States Parties still being bound by the provisions of the present Convention and any earlier amendments which they have accepted.

Article 51

1. The Secretary-General of the United Nations shall receive and circulate to all States the text of reservations made by States at the time of ratification or accession.

2. A reservation incompatible with the object and purpose of the present Convention shall not be permitted.

3. Reservations may be withdrawn at any time by notification to that effect addressed to the Secretary-General of the United Nations, who shall then inform all States. Such notification shall take effect on the date on which it is received by the Secretary-General.

Article 52

A State Party may denounce the present Convention by written notification to the Secretary-General of the United Nations. Denunciation becomes effective one year after the date of receipt of the notification by the Secretary-General.

Article 53

The Secretary-General of the United Nations is designated as the depositary of the present Convention.

Article 54

The original of the present Convention, of which the Arabic, Chinese, English, French, Russian and Spanish texts are equally authentic, shall be deposited with the Secretary-General of the United Nations.

In witness thereof the undersigned plenipotentiaries, being duly authorized thereto by their respective Governments, have signed the present Convention.